Representing Animals

REPRESENTING ANIMALS
is Volume 26 in the series

Theories of Contemporary Culture
Center for 21st Century Studies
University of Wisconsin–Milwaukee

Daniel J. Sherman
General Editor

NIGEL ROTHFELS EDITOR

Representing Animals

INDIANA
University Press

Bloomington & Indianapolis

Part title illustrations by Lisa Moline

This book is a publication of

Indiana University Press
601 North Morton Street
Bloomington, IN 47404-3797 USA

http://iupress.indiana.edu

Telephone orders 800-842-6796
Fax orders 812-855-7931
Orders by e-mail iuporder@indiana.edu

The paper used in this publication meets the minimum requirements of American National Standard for Information Sciences—Permanence of Paper for Printed Library Materials, ANSI Z39.48-1984.

Manufactured in the United States of America

Library of Congress Cataloging-in-Publication Data

Representing animals / edited by Nigel Rothfels.
    p. cm. — (Theories of contemporary culture ; v. 26)
Includes bibliographical references and index.
  ISBN 0-253-34154-X (cloth : alk. paper) — ISBN 0-253-21551-X (pbk. : alk. paper)
  1. Human-animal relationships. 2. Animals—Psychological aspects. I. Rothfels, Nigel. II. Series.
  QL85 .R46 2002
  306.4—dc21

2002004041

1 2 3 4 5   07 06 05 04 03 02

Contents

Introduction
Nigel Rothfels

In the spring of 2000, a strange anniversary spawned renewed interest in an animal that had come to inhabit the human subconscious as much as any physical environment. The occasion was the twenty-fifth anniversary of the opening of the film *Jaws,* and news media around the world turned their attention to Peter Benchley, the story's creator. I want to take a moment here in the introduction to this volume to examine the phenomenon of *Jaws* as a window into human expectations of our relationship with animals. This case is revealing because it demonstrates the deep connections between our imagining of animals and our cultural environment. Furthermore, as the essays in this book illustrate in far greater detail, our cultural environment is rooted in a wide range of circumstances, each with unique but interrelated historical roots.

To mark the film's anniversary, the National Geographic Society (NGS) asked Benchley to return to the creature with which he had so successfully tapped into deep fears of oceans, predators, and monsters. Benchley ventured to the southern coasts of Australia and Africa in pursuit of great whites. From these excursions came the first full-length article on the creature for *National Geographic Magazine,* a special NGS film, several videos, a feature group of pages on the NGS website, interviews by major news services, and at least one on-line chat. On top of all this, an anniversary edition of the film was simultaneously released on DVD. This expanded version included retrospective comments on the making of *Jaws* by, among others, Benchley, director Steven Spielberg, producers David Brown and Richard Zanuck, actors Richard Dreyfuss and Roy Scheider, and the creators of the shark effects.

Part of the continuing interest in *Jaws* can be attributed to the fact that the film has never been forgotten by a certain, commercially significant generation. Nevertheless, what made all this a compelling story was that, over the course of twenty-five years, Benchley had evidently faced his demon and concluded that it was not quite the horror he had imagined. Indeed, concerned that world populations of all species of sharks were declining due to hunting and indifference, Benchley had become a shark advocate. As he succinctly put it, "I couldn't possibly write *Jaws* today . . . not in good conscience anyway." News articles carried headlines like "*Jaws* Author Teams with Sharks" and suggestive, if slightly inaccurate, leads such as "Peter Benchley, author of the 1974 blockbuster *Jaws,* says if he had the book and movie to do over again, he wouldn't," showing that the story had caught the attention of editors, at least. When one reads further into the bodies of the articles, of course, it is clear that Benchley is far from wishing he had never written *Jaws;* his more simple point is that he would never try to write that *kind* of story today.[1] In an interview for the Associated Press, for example, he explains, "I wouldn't do it at all. I wouldn't try to demonize an animal." According to Benchley, "[s]ociety has changed, [and] the perceptions of animals have changed," and to write a book today in which "an animal is a conscious villain is not acceptable anymore. I don't mean acceptable from the political correctness point of view, I mean acceptable morally and ethically and every other way" ("Author Benchley Talks").

Tales of the hunter-turned-conservationist, the foe-turned-friend, are nothing new, and Benchley's work for NGS has all the usual features. We meet other comrades in the struggle to save the sharks—figures like Rodney Fox, who, after barely surviving an attack in 1963 that resulted in 462 stitches, has "ever since devoted his life to the study and protection of great white sharks" (Benchley 12).[2] We are asked to sympathize with the new enthusiast as he grapples with the task of defending something which occasionally eats people. We are asked, in short, to put aside our stereotypes and prejudices and reexamine why we hate or fear sharks. Central to the great white's recuperation are the sometimes contradictory efforts to encourage us both to see the animal in all its magnificent and deadly beauty and to attempt to see it again with the innocence and wonder of a child. Standing beside a huge female great white that had drowned wrapped up in a "longline" pulled by commercial fishermen, Benchley is overwhelmed by the fascination of the twelve thousand people who have gathered to see her dissection in a small South Australian town north of Adelaide. In his article for *National Geographic,* Benchley tried to capture the moment, which, he admits in the film *Great White Shark: Truth behind the Legend,* finally gave him a new way of talking about the beast—the beast that everyone thought they already knew. Alongside a photograph of a child touching a tooth in the mouth of the shark while another child, perhaps three years old, reaches out to lay her hand on the creature, Benchley writes,

> To be sure, she was impressive: about 18 feet long, 3,000 pounds, a robust, mature female with teeth two inches long and dark, impenetrable eyes. Child after child, adult after adult touched the shark not only with their fingertips but with their entire hands, as if to commune with the great creature. They were not afraid; they were awed, almost reverent. (14)

This is a scene, Benchley believes, that was simply unimaginable twenty-five years ago. At that time, he argues, "it was OK to demonize an animal, especially a shark, because man had done so since the beginning of time, and, besides, sharks appeared to be infinite in number" (12).

From the very beginning of the *National Geographic* article, its mission is clear. In large, boldface type on the article's foldout cover pages, the editors write, "Twenty-five years after *Jaws* terrified moviegoers, author Peter Benchley and photographer David Doubilet portray the sea's largest predatory shark in a different light." Far from the foreboding menace-from-below signaled by John Williams's unforgettable soundtrack, this was to be a new great white. No longer the "eating machine" of "fantasy," this new creature, to echo the words of Andrew Isenberg in this volume, was now to be understood as perhaps the most important figure in the ocean's "moral ecology"—the popular idea that Nature is, or has evolved into, a sacred and stable system which has been repeatedly and almost always disastrously disrupted by man, whose solemn obligation has become to try to restore Nature to its *status quo ante* and preserve it in that state for the future. As Benchley puts it in the very last lines of his essay, and in ways familiar to us all at this point,

> Great white sharks have survived, virtually unchanged, for millions of years. They are as highly evolved, as perfectly in tune with their environment as any living thing on the planet. For them to be driven to extinction by man, a relative newcomer, would be more than an ecological tragedy; it would be a moral travesty. (27)

Clearly, behind a great deal of the impetus for rethinking the importance and splendor of the great white—perhaps our equivalent of another pelagic great white from over a century ago—is the hope that the new portrayals of arguably the most enigmatic and charismatic of the ocean's predators will help all sharks before they become only the pathetic, ever-smaller side-catch of commercial fishermen and the dying offal cast overboard after the harvest of yet one more dorsal fin for shark-fin soup.[3]

Reading Benchley's article and his interviews, watching the NGS videos of great whites with their only occasionally ominous soundtracks, it does, in fact, seem that our ideas of sharks are changing. Indeed, I suspect most people would forgive the more obvious excesses of Benchley's (or NGS's) melodramatic tone —including the tendencies to ennoble the biological significance of the great white and to elevate the ethical significance of certain kinds of human involvement in its future survival or demise. Keeping in mind that the almost endless longlines used by commercial ocean fishers do indiscriminately kill thousands of non-targeted creatures, one could, and perhaps should, argue that epic tales are appropriate to awaken people to the historic changes occurring in the world's environments.

With that said, however, there are at least two other significant aspects of NGS's great white shark materials that should give shark advocates—including Benchley—reasons for concern. The first and most conspicuous is that, however much Benchley and others may be interested in putting forth a new image of great white sharks, the producers of the magazine and the videos take every opportunity to promote an older and quite familiar image. While both the opening double-foldout photograph and the closing double-page photo of the *National Geographic* article, for example, feature an open abyssal great white mouth inches in front of the camera, the back-cover text of the video documenting the making of the article struggles to get beyond the creature-from-hell vocabulary of the past:

> *Uncover the truth about Great White Sharks.* Mythic, monstrous . . . misunderstood. Now, on the 25th anniversary of the hit film *Jaws,* National Geographic

embarks on a voyage of breakthrough discoveries that will forever alter our view of the infamous creature once labeled "the perfect killing machine." . . . Witness amazing great white shark behaviors, including never-before-seen footage of explosive, air-borne "breaching" attacks. Experience the breathless fear of a diver trapped on the sea bottom for over four hours by a hungry great white. . . . With cinematography that takes you literally into the gaping mouth of this fearsome killer, it's a surprising new look at one of the largest, most fascinating predators ever to swim the seas—or haunt our imaginations. (*Great White Shark*)

The burden of most of the film, in fact, turns on the struggles of the still photographer, David Doubilet, as he tries to capture "absolute sharkness." Not surprisingly, that quality is sought only in the great white's savage biting, and with any luck savagely biting the camera itself. In short, if Benchley is concerned that the older demonic image of the great white has compounded the difficulties the animal has faced over the last quarter century, his article and the related NGS materials do little to make the animal somehow more sympathetic.

A more subtle but no less significant problem with the NGS projects stems from the basic conceit of most nature films that no one (much less an extensive crew) stands behind the camera and that what we see before the camera is an unmediated, unedited experience of "Nature." These films are highly constructed endeavors in which, among other things, camera angles and exposures are carefully worked out in advance; animals are enticed, coerced, or otherwise manipulated into becoming performers; and overall storylines are fashioned to meet specific, conventional narrative expectations. (Such expectations include, for example, quests in search of a truth which is only and somehow miraculously reached near the end of the film.) The ethical dubiousness of this kind of work by "advocates" for "wild" animals is, of course, virtually never touched upon in the magazines, videos, and cable programs helping to create a culture in which tourists go to Africa to see the Discovery Channel live.

Indeed, in what might have been an enlightening moment in the Benchley article, in a section focusing on the breaching attacks of great whites off the coast of South Africa, the author and NGS quite typically avoid considering themselves as yet one more of the predicaments facing the sharks. Benchley introduces a "young doctoral candidate" named Rocky Strong who had joined Benchley's team for a few days while they filmed the breachings by pulling a piece of seal-shaped plastic with a "videotape camera in its belly." Amazed by the behaviors he was seeing, according to Benchley, "Rocky did, however, find the breaching to be a possible cause for concern. 'Each unsuccessful breach consumes a tremendous amount of energy,' he said. 'If there's a lot of debris on the surface, these guys could conceivably wipe themselves out chasing shadows'" (Benchley 27). The question of whether these guys might wipe themselves out chasing plastic seals is simply never posed.[4]

Perhaps, then, our ideas of sharks have not changed that much in twenty-five years. But they have changed some, and this fact brings me to the substantial reason why I have focused this introduction on Peter Benchley, the anniversary of *Jaws*, and the NGS. In talking about his changing ideas of sharks, Benchley adopted a historian's position, noting that over the last twenty-five years "[s]ociety has changed" and "perceptions of animals have changed." Clearly, while a great many of us may have been terrified by *Jaws* as teenagers in 1975, teenagers today are likely to find the film more quaintly amusing than anything else. In short, we are all pretty comfortable believing that people in 1975, 1950, 1890, or even 1789 had strange, and perhaps unenlightened, ideas about various kinds of animals. With this said, many of us would also feel fairly comfortable with the expectation that society will change again in another twenty-five years and that our perceptions of animals may also change. The problem with this way of thinking is that we end up having to accept that our current, scientific, heavily researched ideas about animals are in a state of constant transformation and that we do not really know what we think we know about them. By this way of thinking, what Jane Goodall, for example, has learned about chimpanzees is mostly just a reflection of broader cultural preoccupations expressed in all kinds of different venues over the last four decades. In a sense, her discoveries are as much about humans as about chimpanzees, and this is a point she might happily accept, though probably for different reasons.

The essays in this volume are part of an accelerating scholarly interest in animals and their place both within and outside of human cultures. Indeed, debates about the significance and representation of animals have become an almost constant presence in our culture. Media coverage of, for example, the shark attacks off the Atlantic coast of the United States in the summer of 2001 (Peltier; McCarthy; "Boy Dies") and the one-eyed-lion Marjan and other animal victims of the war in Afghanistan (George; "Kabul's One-Eyed Lion"; and "AZA Members"), and the perennial stories of whales, elephants, pandas, and other charismatic species, make clear that the stakes in representing animals can be very high. Who controls that representation and to what ends it will be used will be of profound importance in coming years as arguments over global climate change, disappearing and disfigured frogs, razed rainforests, hunting rights, fishing stocks, and the precedence of human needs continue to build.

The idea that the way we talk or write about animals, photograph animals, think about animals, imagine animals—represent animals—is in some very important way deeply connected to our cultural environment, and that this cultural environment is rooted in a history, forms the fundamental basis of this volume. But while all the essays in this book share a common interest in tracing or exploring the ways people have thought about or presented animals in different cultural and historical circumstances, they have been grouped according to shared methodological or thematic interests. The first group considers ways animals have been imagined within discrete historical settings, the second explores different theoretical approaches to understanding the animal object, and the third looks at a series of contemporary settings for human representations of animals.

In the volume as a whole, the issues addressed vary widely. Erica Fudge begins the discussion by providing a historiographical introduction to the topic of animals while asking vital questions about what might constitute a history of animals. Kathleen Kete engages the ideologies that have been ascribed to animals in shifting historical contexts, particularly with respect to animal protection. In her essay on the unique cultural notion of "dog years," Teresa Mangum examines Victorian fiction as a source of our enduring identification with the plight of aging, anthropomorphized animals. Andrew Isenberg, meanwhile, focuses on the changing identification of wildlife as an indicator of the "natural" in our world.

Moving from historical to more theoretical concerns, Steve Baker's study of animals in recent art suggests that artists make or remake animals in ways that produce (and sometimes reproduce) very deliberate but ephemeral meanings. Marcus Bullock undertakes an analysis through literature of what animals seem to mean to us when we look at them and they look back at us. Studying photography and technology at the beginning of the twentieth century, Akira Mizuta Lippit then examines how the idea and metaphor of the animal are embedded in the very ideas of technology and industrialization.

Simultaneously historical and theoretical, the last four essays of the book consider animals in contemporary settings which seem both very familiar and yet somehow remote and strange. Garry Marvin's descriptions of foxhunting in modern Britain offer anthropological insights into a human practice that involves animals as both collaborators and quarry. Jane Desmond demonstrates the connections between traditional taxidermy and high-tech animatronics as a case study of how we pose animals to suit our perceptions of them and of nature more broadly. In a study that introduces a recent endeavor in biotechnology, which many readers will probably follow online, Susan McHugh looks at the motivations and promises of creating a new breed of über-pet through cloning. Finally, in my own essay, I examine the relationship with animals we have made for ourselves in zoos, where we expect our experiences to provide an opportunity to be with animals in their natural habitats without leaving the comfort of our familiar cultural space. Different as the specific subjects of these essays may be, they nevertheless resonate with a common idea—an idea suggested by Benchley, but then quickly passed over—that the way animals are understood is bound in time and place, and that the careful scrutiny of that understanding reveals not only important limits to our knowledge of animals but important limits to our knowledge of ourselves.

If Benchley's reexamination of the great white shark seems suitable to suggest the broader thematic interests of the contributors to this volume, however, there is a circumstantial reason why his work occupies so much of this introduction. By a happy coincidence, NGS's film *Great White, Deep Trouble* was broadcast on Saturday, April 15, 2000,[5] and the weeks preceding the showing saw a flurry of publicity about Benchley and sharks. These were also the weeks leading up to a conference on "Representing Animals," which I co-organized with Andrew Isenberg and which ran April 13–15. My work on this volume began with that conference, and I would like to acknowledge the people who served as resources in various capacities and thank them here. The Center for Twentieth Century Studies at the University of Wisconsin–Milwaukee sponsored the conference as the final event in a year of research at the Center dedicated to the way people think about animals, and most of the essays in this volume were presented in shorter form at that time. Over the course of those three days, thirty-six speakers presented their work. The conference was exciting and challenging for all of the participants, as the Center provided the remarkable intellectual setting—including a public lecture by Dr. Jane Goodall and a building-covering installation by artists Lane Hall and Lisa Moline—for which it is justly well known.

Neither the conference nor this volume would have come about without the support of the Center, now renamed the Center for 21st Century Studies, and its past director Dr. Kathleen Woodward. I had the very good fortune of being hired as an editor at the Center in 1994. Over the course of six years, the Center provided me with a unique intellectual home and I was honored when Kathleen and the Center Advisory Board endorsed my proposal to dedicate a year of the Center's resources to the topic of animals. To give such substantial support to a proposal originating from outside the faculty was an unprecedented move at our university, but the decision demonstrated once again the progressive thinking that the Center has always fostered. It is my pleasure to have the opportunity to thank also the Center's Executive Director and Managing Editor, Dr. Carol Tennessen, who, as everyone who has come into contact with the Center has always known, plays a central role in making everything happen. I also owe many thanks to Dr. Christian Young, the Center's current Assistant Director and Associate Editor. Chris joined the Center as this volume was entering its final stages, took the project in hand, and, with his deep familiarity with the material and his deft ability to keep the contributors on track, brought it to its conclusion. Thanks are also due to Dr. Kristie Hamilton, the Center's Interim Director, who has remained committed to this project while being pulled in every direction as the Center passed though an important transitional period. Finally, my thanks to Patti Sander and Maria Liesegang, past and present Center Business Managers, and to Jason Brame, Keith Chevalier, Ted Wesp, and Terri Williams, graduate student assistants at the Center, for all their good humor, wise advice, and hard work through both the conference and the preparation of this collection.

Notes

1. In an on-line chat Benchley clarified that he does not regret writing *Jaws;* he insisted only that "with the knowledge that we have today *Jaws* would be impossible. Any story about an animal that I would write today would have to portray the animal as the victim, not the villain" ("National Geographic Chat," response to nigeltr-guest).
2. Unlike the article for *National Geographic Magazine,* the video of the "story

behind the story" makes it clear that after recovering from the attack, Fox actually went after sharks for something like blood-vengeance, using explosive-tipped spears that would blow their heads off (*Great White Shark*).

3.   Benchley also toured Asia to advocate for sharks and against shark-fin soup. See "'Jaws' Author Derides."

4.   Nor is the potentially problematic reliance of the team on the expertise and techniques of commercial operators running sharkseeing excursions ever explored in these materials.

5.   *Great White, Deep Trouble* aired on National Geographic's *Explorer* program on CNBC in the U.S. This is presumably the same show packaged as *Great White Shark: Truth behind the Legend*. Another related NGS film taken from the same materials is *Hunt for the Great White Shark,* which focuses on Rodney Fox. There was also a special exhibit in Explorers Hall at the National Geographic Society called "Great White Sharks: In Danger Down Under," which featured an eighteen-foot shark and a shark cage in which visitors could have their pictures taken.

Bibliography

"Author Benchley Talks about 'Jaws.'" *Milwaukee Journal Sentinel Online,* 29 Mar. 2000. Accessed 22 Jan. 2002 <http://www.jsonline.com/news/nat/ap/mar00/ap-benchley-sharks032900.asp>.

"AZA Members to Help Struggling Kabul Zoo." Press Release. *American Zoo and Aquarium Association,* 27 Nov. 2001. Accessed 22 Jan. 2002 <http://www.aza.org/Newsroom/PRkabulzooAZA/>.

Benchley, Peter. "Great White Sharks." *National Geographic,* Apr. 2000: 2–29.

"Boy Dies after Shark Attack." *CNN.com,* 2 Sep. 2001. Accessed 22 Jan. 2002 <http://www.cnn.com/2001/US/09/02/shark.attack/>.

Bredar, John. "Great White Shark." *National Geographic.Com,* Apr. 2000. Field Tale #X77392. Accessed 22 Jan. 2002 <http://www.nationalgeographic.com/fieldtales/greatwhite>.

Dudek, Duane. "'Jaws,' 'Independence Day' DVDs Celebrate Movie Season." *Milwaukee Journal Sentinel Online,* 29 June 2000. Accessed 22 Jan. 2002 <http://www.jsonline.com/Enter/homeenter/videos/jun00/video30062900.asp>.

George, Marcus. "Animal Victims of Afghan Conflict." *BBC News Online,* 6 Dec. 2001. Accessed 22 Jan. 2002 <http://news.bbc.co.uk/hi/english/world/south_asia/newsid_1695000/1695169.stm>.

Great White Shark: Truth behind the Legend. National Geographic Society film, 2000.

"Jaws Author Derides Shark Fin Soup." *Milwaukee Journal Sentinel Online,* 20 July 2000. Accessed 22 Jan. 2002 <http://www.jsonline.com/Enter/gen/ap/jul00/ap-jaws072000.asp>.

"'Jaws' Author Seeks To Help Sharks." *Milwaukee Journal Sentinel Online,* 8 July 2000. Accessed 22 Jan. 2002 <http://www.jsonline.com/Enter/gen/ap/jul00/ap-jaws070800.asp>.

"'Jaws' Author Teams with Sharks." *Milwaukee Journal Sentinel Online,* 5 Apr. 2000. Accessed 22 Jan. 2002 <http://www.jsonline.com/Enter/tvradio/ap/apr00/ap-ap-on-tv-jaws040500.asp>.

"Kabul's One-Eyed Lion Soldiers on." *BBC News Online,* 21 Nov. 2001. Accessed 22

Jan. 2002 <http://news.bbc.co.uk/hi/english/world/south_asia/
newsid_1666000/1666987.stm>.

McCarthy, Terry. "Why Can't We Be Friends?" *Time.com.* Accessed 22 Jan. 2002
<http://www.time.com/time/2001/sharks/cover.html>.

"National Geographic Chat with Peter Benchley." Transcript. *National Geographic.Com,*
Apr. 2000. Accessed 25 Feb. 2002 <http://www.nationalgeographic.com/tv/
explorer/chats/benchley_transcript.html>.

Peltier, Michael. "Florida Bull Shark Swallows Boy's Arm." *ESPN Outdoors.* Accessed
22 Jan. 2002 <http://espn.go.com/outdoors/general/news/2001/0710/
1224495.html>.

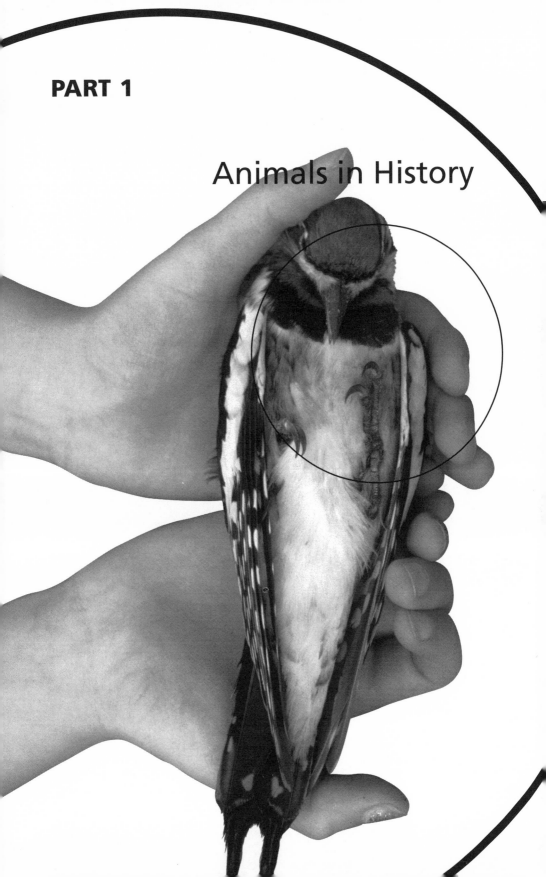

PART 1

Animals in History

1　A Left-Handed Blow: Writing the History of Animals

Erica Fudge

In 1940, Walter Benjamin wrote that "every image of the past that is not recognized by the present as one of its own concerns threatens to disappear irretrievably" ("Theses," 247). The implications of Benjamin's statement are twofold. First, there are elements of the past that are deemed unworthy of entry into conventional history, and it is the obligation of the radical historian to ensure a place for these elements. Second, if that past is allowed to disappear it will take with it a knowledge of the present, because the two are inseparable. In fact, history is where both the past and the present must be brought together, and the historian has a duty to both.

Benjamin is not alone in his sense of the importance of the work of the historian. Just a brief look at the opening statements of two of the most important journals within the discipline underline this fact. In February 1952, the original editors of *Past and Present* wrote, "[H]istory cannot logically separate the past from the present and the future," and they quoted Polybius's idea that the study of history allows us "to face coming events with confidence" (iii). Again, the distinction of then and now, past and present, is refused, and history is figured as a project not merely of recollection, but also of future planning.

From a very different perspective, the founding editorial collective of *History Workshop: A Journal of Socialist Historians* (now called *History Workshop Journal*) argued, "We believe that history is a source of inspiration and understanding, furnishing not only the means of interpreting the past but also the best critical vantage point from which to view the present" (2). And in a following editorial on "Feminist History," Sally Alexander and Anna Davin made the case for the "use" of history—its role as a project of the past, but *for* the future. "Sexual divisions are being questioned now because of the women's liberation movement, and it is through investigating the problems which feminism has raised that we can expect the most useful women's history to emerge" (4–5). All three perspectives—Benjamin's, *Past and Present*'s, and *History Workshop*'s—emphasize the role that history can—and should—play in contemporary culture.

A history of animals would seem to be an obvious place where yet again the ethical nature of the historian's work should be clear. Just as Alexander and

Davin emphasize the formative role of the women's liberation movement in the work of women's history, so it is impossible not to link the recent emergence of histories of animals to the growing centrality of debates about animal rights and welfare. Some histories, for example Richard Ryder's *Animal Revolution: Changing Attitudes towards Speciesism*, have been written that are directly a part of the project of animal liberation. In this book Ryder, conventionally, claims a use for his work: "[S]pecies alone," he writes in his introduction, "is not a valid criterion for cruel discrimination" (6), and as the rest of his work shows, "the motives for speciesist exploitation are multiple. . . . [A]ll are culturally shaped" (333). The book's aim is to explore the ways in which culture shapes our current attitudes, with the intention of changing them. But other histories of animals are also emerging that seem to be less directly linked to what might conveniently be called "activism," and it is these works that I am interested in here. What is the ethical work performed by a history of animals that might appear on the surface to be just another aspect of human history? This essay is an attempt to trace an answer to this question by exploring some of the historiographical issues thrown up by the entry of animals into the arena of history, and it is an attempt to outline how future work might reflect current ethical concerns.

As I began this work, one of the things that immediately struck me was the fact that very little has been written about the historiographical issues raised in writing about animals, and that, in fact, one of the most extended discussions of this topic is probably a joke. The article to which I refer comes from 1974 and is about the need for a history of pets. Published in the *Journal of Social History,* Charles Phineas's "Household Pets and Urban Alienation" is a parody of that rising star of the historical firmament—social history.[1] It is worth quoting at length to give a sense of its argument:

> It seems brash to suggest that pets become the next 'fad' subject in social history, but, after running through various ethnic groups (and now women) historians may need a new toy. There are other promising possibilities. Homosexuals deserve a history, but a movement in this direction has not materialized, perhaps because homosexuals lack political muscle, perhaps because of more personal tensions among historians. Left-handers, another large group long subject to intense social discrimination, merit attention, but again their collective consciousness has lagged. So why not pets? Here, clearly, would be the ultimate history of the inarticulate. Written records, where available, would lend themselves more to anal than to oral history, and a new field could open up. Yet it may not come to pass. Without political power or claims, pets will hardly attract the interest of radical social historians. And at the other pole of academe, university administrators will be under no pressure to add courses on the history of pets, until such time at least as obedience schools are merged with standard undergraduate fare. (339)

The article notes developments within the discipline of history—the emergence of histories of previously unnoticed groups—and takes them to their logical terminus, the history of the most unnoticed of all: animals.

At its heart, the article is a parody of the kind of social, Marxist history that can be traced in the work of a historian such as E. P. Thompson. The pet is placed within the discourse of social history, and its existence within urban culture is given the Marxist spin. Phineas writes, "While granting that escape and particularly the formation of collective protest, as in roving bands of wild dogs, were not common occurrences, it is here that the history of pets should be pursued" (340). This statement is surely a parody of Thompson's documentation of the lives of the workers in *The Making of the English Working Class*. In fact, Phineas is apparently using the kind of social history epitomized in that book to rescue pets from what Thompson infamously termed "the enormous condescension of posterity" (12). "Every gesture of deference," writes Phineas, "every sign of affection among pets was matched by barely-veiled contempt, beginning with resistance to housebreaking" (343). As with Thompson's worker, whose rebellions were sometimes, as he notes, "backward-looking" (12), so the pet could only rebel through a return to the most basic of actions.

Phineas's article, then, is an attack on social history, but it is also strangely prophetic in its recognition of possible developments within the discipline: the history of homosexuality is currently being debated; and, of course, the history of animals is now emerging.[2] One of the arguments of this paper is that the history of animals is not merely a "fad" in the ever widening reach of historical scholarship. Rather, it is a development of existing debates in the discipline as well as in the wider world of human relationships with nature. More than this, I want to argue that the history of animals is a necessary part of our reconceptualization of ourselves as human.

I

There are problems, however, with the idea of a history of animals, and it is worth dwelling on them briefly, before moving on to think about their implications for the development of this area of research. Phineas, albeit ironically, pointed out one of the fundamental issues that faces historians: animals are "inarticulate"; they do not leave documents. Gwyn Prins has noted the traditional belief that "until there are documents, there can be no proper history" (114). It is from the written word that our knowledge of the past comes. Prins, an oral historian, has reason to question this idea and argues that spoken texts continue to be more central than written ones in political as well as popular culture. However, the historian of animals has no such argument available to her: a dog can bark, and that bark can be recorded, documented, but it cannot be understood. The only documents available to the historian in any field are documents written, or spoken, by humans.

Another problem for the history of animals emerges in the ways in which we organize the past in our histories. This is a problem which exacerbates one that is recognized in other fields of history. In 1977, Joan Kelly famously asked the question "Did Women Have a Renaissance?" She answered it by arguing that the term being used to epitomize a historical period actually represented what happened to only a tiny minority of literate men; that it immediately evacuated from the interest of the historian those who were not involved in the intellectual debates—women, the poor, the illiterate. Likewise, in histories of the non-European, a similar question of periodization emerges: L. C. Van Leur, a historian of Indonesia, wondered whether the categories which organize European history—such as "the eighteenth century"—were useful. As Henk Wesseling notes, Van Leur "concludes that there was no point in this since none of the great changes that typify European history of this period can be traced in the Indonesian past" (74). Animals, as far as we know (and this is the only perspective available to us) have no sense of periodization. So, given the question "did dogs have a Renaissance?" the answer is clearly no; dogs did not partake of the intellectual debates which define the period, nor did they have the concept of historical periodization so central to our understanding of the past. If we are to write the history of animals, a wholly different organizing structure would seem to be necessary.

When we take just these two points—the lack of documents and the need for new temporal organization in a history of animals—the whole project becomes rather difficult, not to say impossible. We are attempting to write histories without some of the fundamental ingredients for history. But, as someone with an investment in the history of animals, I do not, of course, wish to declare the project to be futile. The problems I am raising—which, I recognize, sound like they could have been raised by Charles Phineas—are problems which, like the joke article, force us to rethink some of the things that we have perhaps taken for granted. So we must ask another question: is there really an emerging field of work which can be called the history of animals? My answer to this question is both yes and no. The emerging field—containing much absolutely fascinating and rewarding work—is clearly there, but it is not the history of animals; such a thing is impossible. Rather, it is the history of human attitudes toward animals. I continue to use the term "history of animals" as if it were, as Derrida has proposed, *sur rature*—under erasure: it is both indispensable and impossible. It sums up an area of study, but cannot define it.

But, if this history of animals is in reality the history of human attitudes toward animals, we are, perhaps, dealing with something that is merely a part of the history of ideas: nothing really new at all. If our only access to animals in the past is through documents written by humans, then we are never looking at the animals, only ever at the representation of the animals by humans. The difference is an important one, and in a sense epitomizes one of the most significant debates currently taking place within the discipline of history itself, between (broadly speaking) empiricism and poststructuralism—that is, between a belief that the past is recoverable to history through an objective analysis of its documents, and a belief that history is constructed (not always-already there for the taking), and that the documents of the past are always-already only representations. The difference between these interpretations affects how historians can know, can understand the past.

So acknowledging the centrality of representation that emerges in the history of animals places it firmly within what I am terming the poststructuralist camp, and this makes a huge difference to the project. It means, in the first instance, that documents come into being; we read humans writing about animals. Representation is always-already inevitable. But it also, and more significantly, means that the real animal can disappear. That is, the emphasis on the material might be abandoned in favor of the purely textual. Roy Porter has argued that this is a particular problem when dealing with the history of a corporeal substance like the body (208), but it is also fundamentally problematic, I think, when dealing with the history of animals. This issue lies at the heart of Coral Lansbury's *The Old Brown Dog.*

In her study of the Edwardian case of the twice-vivisected "Brown Dog," Lansbury charts the different social groups which became involved in the anti-vivisection movement of the time and offers contextual readings of their motivations. The animal, she argues, was understood in this moment merely as a representation of the (human) self. The women protestors saw their sense of degradation in medical treatment and in pornographic writings echoed in the figure of the vivisected animal; the poor interpreted the use of the dog on the laboratory table as replicating the ways in which some hospitals, under the guise of treatment, used low-status humans for experimentation. These perceived links between animals and humans, however, bring with them dangers. Lansbury notes, "The cause of animals was not helped when they were seen as surrogates for women, or workers. . . . If we look at animals and see only the reflection of ourselves, we deny them the reality of their own existence. Then it becomes possible to forget their plight" (188). By reading the animal as a representation—by managing to displace the central reality of the treatment of the dog—these rioters, Lansbury argues, were not necessarily furthering ideas about animal welfare. Rather, they were using the opportunity to think about their own degraded places in society. The dog is a representation of the human; it is not, paradoxically, a dog.

Lansbury's history, however, is about this repression of the real animal, and rightly points in a different direction. Animals are present in most Western cultures for practical use,[3] and it is in *use*—in the material relation with the animal—that representations must be grounded. Concentration on pure representation (if such a thing were possible) would miss this, and it is the job—perhaps even the duty—of the historian of animals to understand and analyze the uses to which animals were put. If we ignore the very real impact of human dominion—whether in meat-eating, sport, work, or any other form—we are ignoring the fundamental role animals have played in the past. A symbolic animal is only a symbol (and therefore to be understood within the study of iconography, poetics) unless it is related to the real. One way of thinking about why an idealist, purely representational history of animals is a problem is by thinking about the intention of the project: the reason for recovering the history of the animal. In order to do this, a brief outline of some current developments in the field might help to trace some of the interests which histories of animals are currently serving. I should state clearly here that I am not looking at the kind of work that is going on in the fields of environmental and evolutionary history, but am concentrating on work that relies upon textual sources that can be broadly termed social and cultural history.

II

Recent histories of animals seem to take up, broadly speaking, one of three possible positions: these I am terming intellectual history, humane history, and holistic history. The first of these positions can be traced in two recent collections on animals in the Middle Ages: Joyce E. Salisbury's *The Medieval World of Nature* and Nona C. Flores's *Animals in the Middle Ages*. Both offer new ways of reading canonical works of the period, and what is at stake is the deeper understanding of an intellectual debate. Flores writes that the essays in her collection "show how animals were used to convey meaning—whether religious or profane—in medieval culture" (xi). What defines these books as "intellectual" rather than humane or holistic can be clearly traced in Salisbury's introductory claim that the essays in her collection "look at one element of nature but yield much larger truths that reveal the medieval mind" (xii). It is this "medieval mind" that is the main object of analysis, and because of this, these histories seem to reproduce the ideas of the period they are recording. They do not necessarily question them, because what is at stake is an assessment of an intellectual position.[4] Another book that would fit into the category of intellectual history is Keith Thomas's *Man and the Natural World*. Its right to be included here is explicit in its subtitle in the English edition: *Changing Attitudes in England, 1500–1800*. Attitudes, not animals, are the focus.

In the second type of history—humane history—we move away from the intellectual realm into an assessment of the lived relation. It is the materiality of the animal that is important here. A good example of this type of history is Robert Malcolmson and Stephanos Mastoris's *The English Pig*.[5] The authors state their claim for the project clearly:

> All human communities have involved animals. While, in a sense, we all know this, and might regard such a statement as self-evident, the history that we read tends not to pay much attention to species other than our own. History, being written by humans, is mostly about humans; and we may sometimes forget how prevalent— indeed, very visibly prevalent—animals were in most earlier societies. (29)

A rather simplistic paraphrase might be: it is worth writing about animals because animals lived in close contact with humans, and we can learn new things about the humans if we look at the animals. This concentration on the human is something that the authors explicitly acknowledge: "[T]his book," they write, "is mostly about people, for pigs have enjoyed little of what can be called an independent existence" (31). That is, humane history looks at the animals as they are depicted in documents that are always written by humans, and which therefore reveal something of the human.

An apparently very different study that can also be placed in the same field of humane history is Hilda Kean's *Animal Rights*. Here Kean traces the ways in which sympathy for animals led to the organization of animal welfare movements and charities in nineteenth- and twentieth-century British culture. Like Malcolmson and Mastoris, Kean uses the human relation with the animal as a way of looking at broader social (that is, human) ideas. It is not accidental that the subtitle of the book is *Political and Social Change in Britain since 1800,* as one of the central issues here is the growth of popular politics. Kean notes that campaigns such as those against vivisection reflected the "growing influence of women and the working class in political and cultural life" (157). The book is as much a study of the significance and power of popular protest, traced through an analysis of attitudes toward cruelty to animals, as it is about the animals themselves.

Both Malcolmson and Mastoris's and Kean's books are, then, important studies of forgotten aspects of social history, and it is through the animal that these are traced. In fact, this is the acknowledged reason for the work. As Kean writes, "When humanitarians rescued stray animals, or deplored the treatment of cattle driven to slaughter, or erected water troughs for thirsty animals, it tells us more about the political and cultural concerns of society at that time than about the plight of animals per se" (11). The recognition that a new understanding of human life can be traced in animals links Malcolmson and Mastoris's and Kean's work with the intellectual histories, but their concentration on the social, the economic, and the political is also the point of difference. Rather than merely tools for the intellect, animals are the site of social change. However, this difference between intellectual and humane history is, in its turn, the thing that links humane history with the third of my categories: holistic history. I am not saying that there is a clear and absolute division between the categories I am setting up. Rather, I am arguing that they represent different trends within what remains a single body of work that I am terming the history of animals. Ultimately, however, the third of my categories—holistic history—is where I believe an interpretation that can work toward ethical change can be found.

The two outstanding contributions to what I am terming holistic history are Harriet Ritvo's *The Animal Estate* and Kathleen Kete's *The Beast in the Boudoir.* In both of these books the representation of the animal is offered as a way of rethinking cultures which have, apparently, been thoroughly ransacked for meaning by historians, and in both cases—as in Malcolmson and Mastoris's and Kean's work—what emerges is a very new picture of the past.

Both Ritvo and Kete make clear the case for their work. Ritvo says that an examination of animals "illuminate[s] the history not only of the relations between people and other species, but also of relations among human groups" (4). And in a similar vein Kete states, "When bourgeois people spoke of their pets, as they loquaciously did, they pointedly spoke also of their times, and above all else of themselves" (2). On one level it could be argued that what Ritvo and Kete have done is recognize that animals can tell us about humans; this is the humane historian's line. But both go further than this, and what is at stake here is the status of the human itself. The idea that meaning can only be made through difference—which emerges in Saussure's linguistic theory—leads to the inevitable conclusion that the human is only ever meaningful when understood in relation to the not-human. This is a particularly useful conceptualization, in that we learn more about humans by understanding what they claimed that they were not: animals.

In *The Animal Estate* Ritvo concentrates on "rhetorical" strategies; that is, on the ways in which discourse "restructur[es] and recreat[es] . . . reality" (5). It is here that the animal is at its most potent, and, paradoxically, its most materially weak. "Animals," she writes, "were uniquely suitable subjects for a rhetoric that both celebrated human power and extended its sway, especially because they concealed this theme at the same time that they expressed it" (6). Again, a look at the subtitle of Ritvo's work offers a clue to its contents: it is *The English and Other Creatures in the Victorian Age,* and here the reliance of human upon animal for its meaning finds its logical end: human (Ritvo uses the word "English" as a synonym to underline the imperialist belief in the lesser humanity of the non-English) relies on not-human for its meaning, and this reliance creates a sense of a loss of status. The human is just one among "other creatures."

For Kathleen Kete, petkeeping offers a crucial way into understanding what she terms "mediocre lives" (1). And it is through an analysis of petkeeping that she recognizes that the human relationship with the animal "describes the fault lines of individualism" (2). Her analysis of the literature surrounding petkeeping—training manuals, newspapers, lectures, pamphlets—offers a new perspective on human relationships in terms of class and gender, but it also outlines the ways in which nineteenth-century Parisian life represented a clash of ideologies, of *ancien régime* and modernity. The terrors of the new culture were being offset by the bourgeois ownership of animals, creatures who came to represent everything that had been lost—cleanliness, order, and rationality. But inscribed in the pet's function is something deeper: it gives access to what Kete terms "the ruins of Enlightenment thought" (138). Her study, in this sense, fulfills, as does Ritvo's, the purpose of the three classes of the history of animals that I have outlined: the intellectual assumptions and social and political ideas are represented, but, as well as these two elements, Kete reveals the centrality of the animal in human understanding of the self, or perhaps I should say the centrality of the animal in the ways in which humans shore up their fragile status.

In holistic history, then, what emerges is the sense that "human" is a category only meaningful in difference; that the innate qualities that are often claimed to define the human—thought, speech, the right to possess private property; what I have called in *Perceiving Animals* qualities of human-ness—are actually only conceivable through animals; that is, they rely on animals for their meaning. The movement from material to rhetorical, from real to discursive animal, that can be traced in Ritvo's and Kete's works is an inevitable response to some of the problems with the history of animals which I have outlined above, and Ritvo and Kete show brilliantly how turning from the material to the rhetorical need not undercut what I see as the ethical impetus of the history of animals. What the move reveals, in fact, is the way in which use cannot be separated from meaning, and what we see is humans undoing their own status even as they claim they are strengthening it. This is where, I think, the power of the history of animals really lies. Recognizing the centrality of the animal in our own understanding of ourselves as human forces us to reassess the place of the human. If we identify the human as neither a given nor a transcendent truth, then intellectual attitudes that leave unquestioned the result of these assumptions—dominion—must themselves be reviewed as not true, but created. Material and rhetorical are linked in their context, and the history that recognizes this can, in turn, force a reassessment of the material through its analysis of the rhetorical strategies of the written record. The inevitable centrality of the human in the history of animals—the reliance upon documents created by humans—need not be regarded as a failing, because if a history of animals is to be distinctive it must offer us what we might call an "interspecies competence";[6] that is, a new way of thinking about and living with animals. Holistic history, in its redrawing of the human, offers us a way of achieving this.

Recognizing the continuing centrality of humans in the history of animals has two consequences that can upset the wider anthropocentric attitudes. The first is a reexamination of the past and a reassessment of the ways in which humans have perceived and treated animals. The second emerges out of the first, and is a new assessment of our own status as "humans."

III

In Thesis VII of his "Theses on the Philosophy of History" Walter Benjamin writes,

> There is no document of civilization which is not at the same time a document of barbarism. And just as such a document is not free of barbarism, barbarism taints also the manner in which it was transmitted from one owner to another. A historical materialist therefore dissociates himself from it as far as possible. He regards it as his task to brush history against the grain. (248)

Benjamin is writing here of human society, but his ideas about taintedness, about the fact that nothing which is used to maintain power is innocent, however it is presented, are also useful in thinking about the ways in which we live with animals now and in the past. Where Benjamin writes of barbarity, I write of anthropocentrism.

Benjamin noted the transmission of the barbarity of the documents claimed by the victors. Even when the barbarity is counteracted, is protested against, it is still barbarism that rules the day; it is still barbarity that is being expressed. In a recent article M. B. McMullan tells a tale that exemplifies the ways in which the problem of the transmission of barbarity which Benjamin has highlighted might be useful for the history of animals. Following a campaign by the Society for the Prevention of Cruelty to Animals (SPCA) and other reformist groups, and based partly on the fear of the spread of rabies, as well as on the sense of the dogs' physiological unsuitability for the work, in 1839 a new law—section

39 of the Metropolitan Police Act—prohibited the use of dogs to pull carts in London. This act would appear to be based upon humanitarian arguments, but the immediate outcome was far from emancipatory: most of the dogs were killed by their owners because they were too expensive to keep as pets. As McMullan notes, "The measure, purporting to be for their benefit, resulted in their slaughter" (39).

Benjamin argues that the barbarity of the document cannot be eradicated by later interpretations, that these are a mere continuation of barbarity under another name. In the case of the dog carts of nineteenth-century London, the prohibition appears to be a document of humane treatment, a recoiling from the cruelty of dominion which had itself been documented by the carts. On closer inspection the act is, however, a continuation of barbarity under a new guise. As Hilda Kean notes, at the heart of nineteenth-century animal welfare campaigns is the middle-class desire not to be able to see cruelty. Frances Maria Thompson, a patron of the Animal Friends' Society, wrote in the 1830s, "The increasing instances of cruelty in our streets have now risen to such a height that it is impossible to go any distance from home without encountering something to wound our feelings" (Kean 60). It is the wound she feels that is of primary importance; the animal often appears to be of only secondary concern, but the result is an increase in anti-cruelty legislation. This is the anthropocentrism that lay beneath the protests against the dog carts. The barbarity was not halted; it was, as Benjamin recognizes, transmitted from generation to generation. What would appear to be a challenge is really only a continuation: the terms of engagement have not changed, and anthropocentrism is countered with further anthropocentrism.[7]

Terry Eagleton has written that "[a]ny attempt to recuperate the past directly, non-violently, will result only in paralysing complicity with it" (44). I want to argue that to begin to write about anthropocentrism, to note its transmission, is perhaps to begin to dissociate oneself from it, to read it "against the grain." It is a refusal of one of anthropocentrism's strengths—its apparent naturalness. But if we merely recognize that the way in which we understand and inhabit our world remains anthropocentric, we are only part of the way there. The next crucial step, Benjamin argues in his "Theses on the Philosophy of History," is to recognize that our "amazement [that we 'still' do such things] is not the beginning of knowledge—unless it is the knowledge that the view of history which gives rise to it is untenable" (249). A history of progress—one which sees an increase in animal welfare in modernity, one which finds the increase in the number of vegetarians enough to mean the crucial economic centrality of the battery farm can be set aside—fails to see that progress is merely a term which disguises change, a disguise that will always leave us amazed at our own cruelty. By implication, the history of animals cannot merely reflect upon past cruelties, lay them bare for examination with the assumption that such laying bare is itself a political gesture. Actually, in Benjamin's terms, this merely shifts barbarism; it does not counter it. So an alternative must be sought, and it is here that the second possibility of the history of animals comes into its own and offers, I

does undercut some of the assumptions that allow
ne. If we recognize that progress might actually be
r from progressive, then we—humans—are forced
tion with animals. If the rise in charitable institu-
e organizations is premised upon attitudes that are
taint of barbarism, then something very new is
k toward a more equitable relation with animals.

cit reference to Marx's original, E. P. Thompson
history. They are part agent, part victim: it is pre-
which distinguishes them from the beasts, which
td. in Poster 4).[8] Mark Poster notes that Thomp-
tested by few historians, "liberal, conservative, or
of a stable subject who makes his/her own mean-
visions within the discipline. This humanism, this
ower to make their own history, is in Thompson's
wer for those who may be perceived to lack it—
en—but it relied, of course, on animals to make

early advocates for and practitioners of history
iscussions of its possibilities have not abandoned
e early work. Jim Sharpe, for example, argued in
allows us to see "that our identity has not been
rime ministers and generals" (37). Again, identity
is something over which even the most powerless have power. This humanist
idea, as Poster has shown, implies that there is a fixed and stable subject—one
the same in the past and the present—and that it is merely the context in which
this subject finds him/herself that alters his/her being. Humans are born free,
you might say, but everywhere they are in chains.

In this interpretation history from below is the history of how humans have
been chained, and how they have challenged their confinement. And the writing
of history itself becomes one of the greatest challenges: it is where the silencing
of the marginal is ended, where "the condescension of posterity" is undone. The
recovery of the lives of those regarded as unimportant and insignificant by traditional history not only gives a broader view of society, it also allows the historian to reclaim, in the name of the people, a significant part of the ideological
apparatus: history itself. But this history remains humanist in a very obvious
and simple way: in response to Francis Fukayama's declaration of the end of
history, most historians, Poster argues, would state that, in fact, "history is a real
sequence of events that will end only with the last gasp of the last human being.
History and humanity are coterminous" (59). Without our ability to create our
own history, Thompson argued, we are not fully human, and without humanity
there can be no history.

This idea of the stability of the human subject has, of course, come under threat in the work of Michel Foucault. What is lost in Foucault's work is the sense of self as an autonomous being. Instead of the transcendent, stable, Cartesian subject there is a self formed only in discourse, under the strategies of power. This decentering of the subject is clearly related to the work of the history of animals, but one issue separates the two. Foucault and many of his followers do not go beyond the human. Strategies of othering are examined, but only in terms of othering humans; the animal is a powerful rhetorical category into which some humans—the mad, the criminal—are placed. Real animals are not the issue. A good example of this can be found in Stephen Greenblatt's new historicist analysis of *The Tempest*. Greenblatt writes, "Language is, after all, one of the crucial ways of distinguishing between men and beasts. . . . Not surprisingly, then, there was some early speculation [by colonialists] that the Indians were subhuman, and thus, among other things, incapable of receiving the true faith" (23). The opposition "men and beasts" slides into the opposition of human and subhuman, and the animal disappears from view.

In the history of animals, however, to question the anthropocentric view of the world—to brush history against the grain—is to challenge the status of the human, which in turn is to throw all sorts of assumptions into question. If we can no longer assume our own status then we can no longer take the status of animals as a given. What was assumed to be natural—human dominion—is revealed instead to be manufactured, that is, ideological. Through anthropocentrism—the recognition that the only vision is the human vision, the only history a human history—we can in fact work against anthropocentrism, make it untenable.

In a recent article, Malcolm Bull has suggested a way of reading Nietzsche that parallels my reading of anthropocentrism. Bull argues that the anti-Nietzsche is to be found not so much in reading *against* Nietzsche as in reading *beyond* him by refusing the status of the Superman, of the master, which Nietzsche "flatteringly offers . . . to anyone" (124). Bull writes, "The act of reading always engages the emotions of readers, and to a large degree the success of any text (or act of reading) depends upon a reader's sympathetic involvement. A significant part of that involvement comes from the reader's identification with individuals or types within the story" (126–27). Within the work of Nietzsche, Bull argues, this identification is with the Superman: we inevitably "read for victory," and this means that Nietzsche is never canceled. Rather, he is demonstrated. "Reading Nietzsche successfully means reading for victory, reading so that we identify ourselves with the goals of the author. In so unscrupulously seeking for ourselves the rewards of the text we become exemplars of the uninhibited will to power" (128). We are not being flattered when Nietzsche addresses his readers as Supermen: "If [they] have mastered his text, [readers] have demonstrated just those qualities of ruthlessness and ambition that qualify them to be 'masters of the earth'" (128–29). Bull argues that the only counter to Nietzsche is to "read like a loser," that is, to align oneself with the herd and not the Superman. In accepting the argument of the text, but turning it against him/herself, readers will be made to feel "powerless and vulnerable," and it is this that will allow them to move beyond the position of mastery that appears to be theirs (130).

Bull argues that the vulnerability experienced when reading like a loser replicates interspecific relations: "Superman is to man, as man is to animal" (133). By refusing to be positioned as Supermen we are inevitably positioning ourselves as animals, and this, for Bull, is a step forward. It is a step beyond Nietzsche rather than a refusal of him: a refusal, he writes, would allow for the continuation of "the position he chose for himself." An opposition that "comes only from within pre-existing traditions" would allow Nietzsche to "live for ever as [his critics'] eschatological nemesis" (123–24).

I am arguing that a similar maneuver is needed within the discipline of history. Where Bull posits the anti-Nietzsche, I am suggesting the anti-humanist. We must abandon the status of the human as it is presented within humanist history; we must read against this. Instead, we need to assert and assess the ways in which "human" is always a category of difference, not substance: the ways "human" always relies upon "animal" for its meaning. By refusing humanism, and, implicitly, anthropocentrism, we place ourselves next to the animals, rather than as the users of the animals, and this opens up a new way of imagining the past, something that has to be central to the project. If it is to impact upon questions about the ways in which we treat animals today, if it is to have something to add to debates about factory farming, cruel sports, fur farms, vivisection, and the numerous other abuses of animals in our cultures, then the history of animals cannot just tell us what has been, what humans thought in the past; it must intervene, make us think again about our past and, most importantly, about ourselves. The history of animals can only work at the expense of the human.

But this is not to say that the fragmentation of the human—its lack of fixity— is the way forward. Rather, I want to suggest that by recognizing the lack of foundation for our perceived stability we can begin to think about the category "human" in very different terms. History and humanity are, as the humanists proclaim, coterminous, but a history can be written that does not celebrate the stability of what was, what is, and what shall be. Instead history should reinterpret the documents of the past in order to offer a new idea of the human. No longer separate, in splendid isolation, humans must be shown to be embedded within and reliant upon the natural order.

Wendy Wheeler has discussed the need for change in our political, emotional, and working lives, and her ideas are useful here. She writes of a modernity in which melancholia, which is "characterized by punitive and vicious self-loathing, and by an inability to let go and move on," is the organizing principle of our world, and she argues for a new modernity in which mourning—which allows the individual to "transform the shattered fragments of an earlier self and world, and to build something new from those fragments and ruins"—will be the key (165). Using developments in neurobiology, Wheeler claims that this change from melancholy to mourning is based on the decline of the "old cartesian divide," a decline which will give way to "more complex holistic models of both the individual's understanding of the relationship between mind and body and, more widely, the relationship between individual creatures and the living world of which they are a living part." This Wheeler terms an "ecological sensibility" (165).

Likewise, we must write a history which refuses the absolute separation of the species; refuses that which is the silent assumption of humanist history. By rethinking our past—reading it for the animals as well as the humans—we can begin a process that will only come to fruition when the meaning of "human" is no longer understood in *opposition* to "animal." Then "human" can be recognized as meaning something quite new: a being which only differentiates itself by being able to write and interpret its own history. If this is so, it is only right that we should ensure that this history is the one we deserve.

In his spoof article Charles Phineas likened the need for the history of pets to the need for the history of the left-handed. This connection between the human "other" and the animal is not a new one, but his connection has a pleasing resonance with a statement by Walter Benjamin. In "One Way Street" Benjamin wrote, "All the decisive blows are struck left-handed" (65). From the most unexpected place comes the most disruptive assault. The history of animals has the potential to be such a left-handed blow to many of the anthropocentric assumptions we have about ourselves. And by this means it can become, I think, a powerful part of our revisioning of our place in the world.

Notes

I am grateful to Clare Palmer, Wendy Wheeler, and Sue Wiseman for reading and commenting on an earlier draft of this essay.

1. I am extremely grateful to Anne Goldgar for pointing out this article to me. While researching this paper I contacted the editor of the *Journal of Social History* in an attempt to discover the status of the article, but unfortunately, the journal had no record of Charles Phineas nor of the nature of his article. However, the previous issue of the *Journal* did contain a spoof article—Diana Shroud, "The Neolithic Revolution: An Analogical Overview." This spoof was acknowledged in an "Editor's Note" in the same number in which Phineas's article appeared (368). So spoofing as a way of raising some interesting historiographical issues was certainly a part of the work of the *Journal* at the time Phineas's article appeared.

2. Peter Burke, writing in 1992, seems to have recognized the emerging reality of this turn to nature within history when he noted, "Today, the very identity of economic history is threatened by a takeover bid from a youthful but ambitious enterprise . . . eco-history" ("Overture," 1).

3. The pet might be considered an important exception, and much has been written on its place in culture (see, for example, Shell).

4. This is not, of course, the only way in which medieval scholars have inter-

preted their period. In her own monograph, *The Beast Within,* for example, Salisbury brings the material and the intellectual positions together brilliantly.

5. In this book Malcolmson and Mastoris do look at the ways in which pigs served a symbolic function in the eighteenth, nineteenth, and twentieth centuries, but the positioning of the chapter "Images of the Pig" at the beginning of the book emphasizes the authors' concentration in the rest of the study on the real, lived relation of humans and pigs in English culture. The symbolic animal serves as a lead-in to the real subject of the book, the real animal.

6. This phrase is an adaptation of Christian Meier's call for an increased "intercultural competence" in European history (34).

7. This anthropocentrism of the animal welfare movement did not end in the nineteenth century. Ted Benton and Simon Redfearn found that some of those involved in the Brightlingsea protests against the live export of veal calves were just as, if not more, interested in their own liberty—their right to protest (51–58).

8. I am indebted to Poster's analysis in the following pages.

Bibliography

Alexander, Sally, and Anna Davin. "Feminist History." *History Workshop: A Journal of Socialist Historians* 1 (1976): 4–6.

Benjamin, Walter. "One Way Street." *Reflections.* Ed. Peter Demetz. Trans. Edmund Jephcott. New York: Schocken, 1978. 61–94.

——. "Theses on the Philosophy of History." *Illuminations.* 1940. Ed. Hannah Arendt. Trans. Harry Zohn. London: Fontana, 1992. 245–55.

Benton, Ted, and Simon Redfearn. "The Politics of Animal Rights—Where Is the Left?" *New Left Review* 215 (1996): 43–58.

Bull, Malcolm. "Where Is the Anti-Nietzsche?" *New Left Review,* 2nd ser., 3 (May–June 2000): 121–45.

Burke, Peter. "Overture: The New History, Its Past and Its Future." Burke, *New Perspectives,* 1–23.

——, ed. *New Perspectives on Historical Writing.* Cambridge: Polity, 1992.

Eagleton, Terry. *Walter Benjamin: Or, Towards a Revolutionary Criticism.* London: Schocken, 1981.

Editorial Collective. "Editorials: History Workshop Journal." *History Workshop: A Journal of Socialist Historians* 1 (1976): 1–3.

Editors. "Introduction." *Past and Present* 1 (1952): i–iv.

Flores, Nona C. *Animals in the Middle Ages: A Book of Essays.* New York: Garland, 1996.

Fudge, Erica. *Perceiving Animals: Humans and Beasts in Early Modern English Culture.* Basingstoke: Macmillan, 2000.

Greenblatt, Stephen J. "Learning to Curse: Aspects of Linguistic Colonialism in the Sixteenth Century." *Learning to Curse: Essays in Early Modern Culture.* Ed. Stephen J. Greenblatt. New York: Routledge, 1990. 16–39.

Kean, Hilda. *Animal Rights: Political and Social Change in Britain since 1800.* London: Reaktion, 1998.

Kelly, Joan. "Did Women Have a Renaissance?" 1977. *Becoming Visible: Women in European History.* 2nd ed. Ed. Renate Bridenthal and Claudia Koonz. Boston: Houghton, 1987. 175–201.

Kete, Kathleen. *The Beast in the Boudoir: Petkeeping in Nineteenth-Century Paris.* Berkeley and Los Angeles: University of California Press, 1994.

Lansbury, Coral. *The Old Brown Dog: Women, Workers, and Vivisection in Edwardian England.* Madison: University of Wisconsin Press, 1985.

Malcolmson, Robert, and Stephanos Mastoris. *The English Pig: A History.* London: Hambledon, 1998.

McMullan, M. B. "The Day the Dogs Died in London." *The London Journal* 23.1 (1998): 32–40.

Meier, Christian. "Scholarship and the Responsibility of the Historian." *The Social Responsibility of the Historian.* Ed. François Bédarida. Special issue of *Diogenes* 168 (1994): 25–39.

Phineas, Charles [pseud.]. "Household Pets and Urban Alienation." *Journal of Social History* 7.3 (1974): 338–43.

Porter, Roy. "History of the Body." Burke, *New Perspectives,* 206–32.

Poster, Mark. *Cultural History and Postmodernity: Disciplinary Readings and Challenges.* New York: Columbia University Press, 1997.

Prins, Gwyn. "Oral History." Burke, *New Perspectives,* 114–39.

Ritvo, Harriet. *The Animal Estate: The English and Other Creatures in the Victorian Age.* London: Penguin, 1990.

Ryder, Richard. *Animal Revolution: Changing Attitudes towards Speciesism.* Oxford: Blackwell, 1989.

Salisbury, Joyce E. *The Beast Within: Animals in the Middle Ages.* New York: Routledge, 1994.

———, ed. *The Medieval World of Nature: A Book of Essays.* New York: Garland, 1993.

Sharpe, Jim. "History from Below." Burke, *New Perspectives,* 24–41.

Shell, Marc. "The Family Pet." *Representations* 15 (summer 1986): 121–53.

Shroud, Diana. "The Neolithic Revolution: An Analogical Overview." *The Journal of Social History* 7.2 (1974): 165–70.

Thomas, Keith. *Man and the Natural World: Changing Attitudes in England, 1500–1800.* London: Penguin, 1984.

Thompson, E. P. *The Making of the English Working Class.* London: Penguin, 1991.

Wesseling, Henk. "Overseas History." Burke, *New Perspectives,* 67–92.

Wheeler, Wendy. *A New Modernity: Change in Science, Literature, and Politics.* London: Lawrence, 1999.

2 Animals and Ideology: The Politics of Animal Protection in Europe

Kathleen Kete

In 1974, the *Journal of Social History* published a spoof on the history of pet-keeping, "Household Pets and Urban Alienation" by "Charles Phineas" of "Boxer University." It was a satire of the kinds of subjects Ph.D. programs were producing in the 1970s, when social historians began to pay attention to the history of everyday life. However, it was published at the moment when a number of books and articles were about to appear which would establish the importance of attitudes toward animals in European, especially British and French, social history. These include Douglas Hay's "Poaching and the Game Laws on Cannock Chase" (1975), James Turner's *Reckoning with the Beast: Animals, Pain, and Humanity in the Victorian Mind* (1980), Maurice Agulhon's "Le sang des bêtes: Le probleme de la protection des animaux en France au XIXème siècle" (1981), P. B. Munsche's *Gentlemen and Poachers: The English Game Laws, 1671–1831* (1981), Brian Harrison's "Animals and the State in Nineteenth-Century England," a chapter of his book *Peaceable Kingdom* (1982), Keith Thomas's *Man and the Natural World: A History of the Modern Sensibility* (1983), Robert Darnton's *The Great Cat Massacre and Other Episodes in French Cultural History* (1984), Robert Delort's *Les animaux ont une histoire* (1984), Coral Lansbury's *The Old Brown Dog: Women, Workers, and Vivisection in Edwardian England* (1985), and Harriet Ritvo's *The Animal Estate: The English and Other Creatures in the Victorian Age* (1987).

The 1980s were notable also for the publication of philosophical and anthropological studies of the place of animals in modern life, including Tom Regan's *The Case for Animal Rights* (1983) (following Peter Singer's *Animal Liberation* of 1975), Yi-fu Tuan's *Dominance and Affection: The Making of Pets* (1984), and Vicki Hearne's *Adam's Task: Calling Animals by Name* (1986). These works were followed in the 1990s by Luc Ferry's *Le nouvel ordre écologique: L'arbre, l'animal, et l'homme* (1992), my own *The Beast in the Boudoir: Petkeeping in Nineteenth-Century Paris* (1994), Harriet Ritvo's *The Platypus and the Mermaid and Other Figments of the Classifying Imagination* (1997), Hilda Kean's *Animal Rights: Political and Social Change in Britain since 1800* (1998), and Erica Fudge's *Perceiv-*

ing Animals: Humans and Beasts in Early Modern English Culture (2000), as well as several studies of the anti-vivisection movement in nineteenth-century Europe.

What do these texts tell us? Is there a narrative of Europeans' relationships to animals from, say, the sixteenth century to the present which would encourage us today in ameliorist thinking? To borrow from a schoolchild's essay in republican France in the 1880s, can we understand the meaning of this narrative to be "we once were cruel to animals, now, after listening to our schoolteachers, we are kind?" (Archives). Does the narrative support our linking of animal liberation to other liberations?

I do not think so. What these texts tell us is that the history of Europeans' relationships to animals can be placed within neither a progressive nor a conservative narrative of history, or, rather, it can be placed sometimes within a left, sometimes within a right political narrative. What is significant is the role the animal/human divide plays in building a sense of social identity in modernizing Europe, in charting a shifting line between an "us" and a "them," a line which unexpectedly runs through the Puritan, bourgeois, feminist, nationalist, and even Nazi revolutions.

In this essay I mean to review for us, who are thinking over the representation of animals at the turn of the twenty-first century, how that representation has been constructed in the past to mark "in" groups and "out" groups, to assert power politically and ideologically. My goal is to challenge any ideas that we still may have that the history of European animal protection is simply, or only, "benign."

I do not mean to recite the entire history of animal protection movements in Europe, but, rather, to concentrate on two regulations. The first is the English Protectorate's ordinance of 1654, which prohibited cockfighting and cockthrowing. The second is the set of Nazi animal protection laws, which were issued immediately upon the Nazi takeover of the German state in 1933 and which became the most comprehensive set of laws protecting animals in Europe.

I discuss the Puritans and the Nazis because each is responsible for a landmark event in the history of animal protection—the first animal protection law, and the most comprehensive set of laws. But more importantly, they lie at the heart of this essay because of their paradigmatic character. The ordinance of 1654 was based on two assumptions which were to last into the twentieth century—one, that traditional behaviors toward animals were socially disruptive, and two, that humans have a duty to be kind to animals, or, at least, to not cause them unnecessary pain. The Nazis worked within a new paradigm. Accepting the logic of modernism, they abolished the line separating human and animal and articulated a new hierarchy based on race, which placed certain species— races—of animals above "races" of humans—eagles and wolves and pigs in the new human/animal hierarchy were placed above Poles and rats and Jews.

I also address the history of hunting, access to which most directly marked the powerful in medieval and modernizing Europe. Monopolies on the protection of game by and for landed elites were broken in Western Europe in the nineteenth century when democratization allowed urban professionals and others access to the killing of game.

My point is not to link animal protection with repression but to de-couple animal protection from the history of social liberation which we intuitively, or Whiggishly, or romantically, wish to uphold. It has been twenty-five years since Peter Singer published *Animal Liberation* and twenty years since the histories of European attitudes toward animals became established in the field. The disjunctions between the two stories which are told, philosophical and historical, are the burden of this essay.

What is the context of Europe's first animal protection law and where does its meaning lie? The Protectorate's ordinance of 1654 was promulgated during the radical Puritan stage of the English Civil War. What exercised the Cromwellian ire were the blood sports of modernizing Europe, cock-throwing and cockfighting, but also dogfighting, bullbaiting, and bull-running, which were played in villages, towns, and fairgrounds—sites associated with drinking, gambling, and brawls.

Cock-throwing was a game traditionally played on Shrove Tuesday and on other festive occasions. The game began with the tethering of a cock to a stake with about a foot or two of slack. The contestants took turns throwing clubs at the cock until it was dead. Bullbaiting was much like cock-throwing. The bull was tethered with a rope long enough to allow it some mobility. Dogs were set upon the bull until it was weakened and bloodied from fighting. The bull then was slaughtered (see Malcolmson 34–51).

Bullbaiting was said to tenderize the meat of male animals. So was bull-running, which took place over an entire day and was a town-wide event. A bull was set loose, then was beaten by people and chased by dogs through the streets of the town. At the end of the day, it was butchered. Traditional recreations merged with the ritual slaughter of animals in the case of cock-throwing, bullbaiting, and bull-running. In each case, these practices began to appear in the historical record as they were legislated against by ascendant social groups.

The Puritans had argued against these practices as early as the mid sixteenth century. Blood sports and other popular recreations were associated with idleness and drunkenness. They profaned the Sabbath. They turned people away from their duties to God. They disrupted the godly society. Moreover, in a profoundly important change in the direction of thinking about the relationship between humans and animals, the Puritan reading of the expulsion of Adam and Eve from the Garden of Eden led to a recognition that humans owed it to animals not to enjoy or increase their suffering, a suffering which had become their lot after Adam's sin. Lords of creation, as revealed in chapter 1 of Genesis—"you will have dominion over the earth and the animals in it"—, the people of the European West began, with the Puritans, to develop instead notions of good stewardship in their treatment of animals (Thomas 150–54).

Puritan opposition to blood sports provoked a counterstatement by the early Stuarts. The King's Declaration of Sports, issued in 1618 and reissued in 1633, was a defense of traditional recreations. Its insistence that traditional recreations—blood sports included—lay outside the purview of reform was repeated by some gentry and some rural poor into the nineteenth century and, with respect to hunting, throughout the twentieth century. The Declaration of Sports helped trigger a Puritan revolt against the state, while anger at Puritan interference in everyday life became a leitmotif of resistance to Puritan revolution.

The ordinance of 1654 was overturned in the Restoration. Middle-class opinion in the next century, however, continued to gather against blood sports. Puritanism resonated with a more generally developing middle-class view which conflicted not just with popular culture, urban and rural, but also with elite culture—the Stuart court and unreconstructed gentry. The valorization of happiness and benevolence expressed in latitudinarianism and more generally in Enlightenment thought was also helping to shape middle-class attitudes toward animals in England. Robert Malcolmson shows how repulsion to these sports was expressed in the municipal press in terms which point us back to the direction of the Puritan ordinance of 1654, a concern with "public order" and an association of kindness to animals with religious or enlightened duty (118–22).

By the end of the eighteenth century, many towns were enforcing ordinances against cock-throwing and bullbaiting. Municipal ordinances were followed in 1835 by the Cruelty to Animals Act, which outlawed the "running, baiting, or fighting" of any animal. Clearly, middle-class reformism was mobilizing, pitting itself against elite and popular conservatism. The Protectorate's ordinance was a key episode in Europe's first social conflict over the treatment of animals. It established a pattern of contention which would continue throughout the modern period.

One important shift in the pattern occurred in Stamford in the 1830s. There, middle-class opinion turned against the abolition of bull-running when the London-based Royal Society for the Prevention of Cruelty to Animals (RSPCA), backed by Royal Army forces, mounted an attack on Stamford's bull-running. The formation and the history of the RSPCA in the nineteenth century will be discussed below. Here we should note that in England lines of conflict over the treatment of animals could be shaped not only by class but also by a divide between state and local traditions in ways which echo the conflict between London and the counties in the age of Civil War and Revolution.

Social conflicts associated with hunting in early modern Europe are also exemplary in that they are distinct again from those formed over the practice of popular blood sports. The rural poor were allied in this issue with urban elites, not in opposition to hunting but in resentment of their exclusion from the sport. The history of hunting reminds us that it is the use of animals by social groups to assert power, as much as it is the development of particular behaviors toward non-humans, which is at issue in the narrative of European attitudes toward animals.

In England and on the continent hunting was reserved for the landed elites. In the Middle Ages hunting was a type of practice warfare for the nobility. By the twelfth century forests were being reserved by important nobles and royalty for hunting. Although game could be a precious source of protein in the premodern economy, historians stress the political and cultural function of hunting. In an age that depended on increasing the arable to expand grain production, the preservation of forests or fragments of forests in deer parks was an exercise of power. The Robin Hood legends indicate the resentments that the royal forest law in England could trigger among those excluded from its benefits.

That hunting was an enduring attribute of monarchy is made clear in the biographies of the early modern monarchs of England and France. Even in old age, Elizabeth I would go shooting. James I liked to bathe his arms in the steaming blood of a dying deer, then anoint the faces of his entourage with its hot blood (Thomas 147, 29). Louis XVI's hunting parties appear frequently in the memoirs of Saint Simon. It is in this context that Louis XVI's journal entry for July 14, 1789, makes sense. Simon Schama explains in *Citizens* that his entry, *rien* ("nothing"), tells us not so much that the king was out of touch with one of the most important of revolutionary events but that he was disappointed at not being able to hunt that day (419). As we will see below, his comment passes for premonition. For three weeks later, on the night of August 4, 1789, the hunting privileges of the noble elite were abolished along with all other aspects of feudalism.

Hunting had been a privilege of the ruling class since the establishment of manorialism. In France the exclusive right of the lord of the manor to hunt on peasants' land was one of the remnants of this system which economic modernity was making obsolete. Tocqueville points out in *The Old Regime and the French Revolution* that this right was, like the *banalité*—which included the obligations of peasants to use the lord's ovens and mills—, less punishing in and of itself than as a reminder of an anachronistic system of power relations (31–32). Hunting's importance in defining social relations in rural France is indicated by the fact that it was both closely guarded by the nobility and contested by the peasantry. Isser Woloch points to the prevalence throughout eighteenth-century Europe of poaching as a form of social protest. He also explains that complaints about the hunting privileges of the nobility were among the most frequent in the *cahiers de doléances,* the lists of grievances solicited by the king on the eve of the French Revolution (177).

Poaching continued in France after the Revolution, redefined as a property crime. The state designed a permit system in the 1830s to combat the problem. But as Eugen Weber notes in *Peasants into Frenchmen,* hunting offenses remained more common than theft in rural areas through much of the century (61). At the same time, however, the romantic tide was turning some of the great landowners against hunting. Witness the lament of the romantic poet and revolutionary Alphonse de Lamartine for a dying deer in his poem "Mon dernier coup de fusil" (see Kete 24).

In the German states and in Russia, where serfdom "hardened" during the eighteenth century, hunting also marked power relations. The obligations of serfs included the beating of game, that is, moving en masse through fields, woods, and underbrush, driving game forward into clearings to be slaughtered by nobles. Readers of *War and Peace* will remember its wolf-hunting scene.

David Blackbourn suggests in *The Long Nineteenth Century* that even in areas where the ties of serfdom were loosest, the hunting rights of the nobility were held fast (5).

For the most part, early modern hunting on the continent was a male pursuit, although, as W. H. Bruford relates in *Germany in the Eighteenth Century,* German ladies were sometimes invited along to pig-stickings (82).

In England, conflicts over hunting were more complicated. Rural capitalism was destroying the medieval manor as urban capitalism was the guilds. By the eighteenth century London was the center of a commercial empire poised to dominate the globe. It is in this context of emerging capitalism that the game laws of early modern England and the opposition they generated can be understood. Though all English game laws were oppressive to the lower classes, historians see the Game Law of 1671 as introducing class conflict into the arena of hunting.

The Game Law of 1671 followed the political logic of the seventeenth century in that it displaced the monarch as sole owner and protector of game by including in that role the landed gentry. The gentry could hunt freely throughout the countryside (subject to a weak law of trespass), and they were charged with protecting game by employing gamekeepers and enforcing the Game Law through their positions as justices of the peace.

For P. B. Munsche, writing in *Gentlemen and Poachers: The English Game Laws, 1671–1831,* it is significant that urban elites—those merchant-capitalist investors in the East and West Indies Companies who had previously joined with the gentry in resisting absolutism—were excluded by the Game Law from hunting. The Game Law qualified only large land owners, not those whose wealth was movable. Munsche argues that the new law must have been aimed at this group, since it did not alter the status of the lower classes with respect to hunting—that is, the penalties for poaching remained the same, a fine of about a day's wages for rural workers (18–19).

In Munsche's view the function of the Game Law was to enhance the social position of the gentry at the expense of the "urban bourgeoisie," held to be responsible for the excesses of the revolution. Merchants were often Dissenters. More vaguely, but importantly, city life was associated with modernity, newness, rootlessness, and change. The importance of hunting in early modern England is that it allowed country gentlemen to build a positive social identity. Their exclusive association with hunting let them assert themselves as simple, natural, and English, a political move that shaped the divide between Tory and Whig in the Hanoverian century.

For Douglas Hay the meaning of the Game Law lies in its enforcement, especially after the mid eighteenth century when amendments made penalties for poaching harsher. Whipping, hard labor, and, by 1800, transportation to Australia for this offense were possible. The killing of deer in a park, that is, in an enclosed area, was punishable by death. Hay analyzed the application of the Game Law on Cannock Chase, a great estate belonging to the Paget family. The law was aggressively enforced through gamekeepers. It was also universally resisted by villagers. Unlike a crime of property, which could alienate the perpetrator from the community, the hunting of game on land once viewed as commons was understood as morally right though legally wrong. Hay shows how villagers protected poachers from Paget's gamekeepers. Poaching, Hay shows, was—like wrecking, smuggling, arson, and rioting—a community crime, a form of protest, a way of building social identity among rural wage workers who were no longer feudal but not yet fully modern and class conscious (see especially 200, 244–46).

In rural England people defined themselves in terms of their relationship to hunting. Hay and Munsche would both agree. For Hay, unlike Munsche, the defining divide was between patricians—gentry and merchants—and plebeians, the working poor of rural and urban England. The Game Law was part of a criminal code, a theater of power, based on the strategic deployment of penalties of capital punishment and transportation, which throughout England maintained the dominance of the propertied over the poor.

In any case, capitalism helped to destroy the Game Law of 1671 and its amendments. Poaching was found to be fueled by the urban elite's demand for game, that is, game poached from the gentry found its way to the urban gullet. The status of game was such that it had become a necessary part of a gentleman's table and of a tavern menu by the early nineteenth century. The Game Reform Law of 1831, which opened hunting to anyone with a permit, was promulgated in part to increase the legal supply of game and make poaching less attractive and lucrative. In this it failed. In its other purpose, however, the law was more successful. Granting access to hunting to doctors, lawyers, civil servants—nineteenth-century young professionals such as one finds hunting through the pages of Anthony Trollope novels—encouraged the adoption of Tory attitudes toward animals as national ones.

In a theme that strengthens as the nineteenth century wears on, Englishness comes to be set apart from other cultures by its special relationship with nature. The democratization of hunting also results in the gentry's finding new ways to express their status with respect to animals. The raising of prize pigs and cows is satirized in the endearing figures of Lord Empworth and his pig in the novels of P. G. Wodehouse and analyzed in *The Animal Estate* by Harriet Ritvo.

In the nineteenth century, attitudes toward animals took on unprecedented political importance. This is true for England especially, where the Society for the Prevention of Cruelty to Animals (renamed in 1840 the Royal Society for the Prevention of Cruelty to Animals) shaped both public opinion and public policy.

Animal protection societies were formed throughout Europe and the United States on the model of the British. The most important European society after the British was the French Société protectrice des animaux (Animal protection society), which was founded in 1845. Societies were also formed in the German states and in Switzerland in the late 1830s and 1840s. The German cities of Dresden, Nuremberg, Berlin, Hamburg, Frankfurt, Munich, and Hanover established societies. In Switzerland, Berne, Basle, Zurich, Lausanne, Lucerne, and

Geneva did so, too. According to Ulrich Tröhler and Andreas-Holger Maehle, a German national organization, the Verband der Tierschutzvereine des Deutschen Reiches (Union of animal protection societies of the German lands), in the early 1880s included more than 150 local animal protection societies. The Swedish national society was founded in 1875.

Marx specifically noted the role of animal protection societies within bourgeois Europe. In *The Communist Manifesto* he grouped them with other humanitarian organizations under the rubric of "Conservative, or Bourgeois Socialism." Marx saw the universalism of bourgeois culture at work in organizations whose object was the reform of lower-class behavior. "[M]embers of societies for the prevention of cruelty to animals," like "temperance fanatics," "organizers of charity," and "improvers of the condition of the working class," Marx wrote, "wish for a bourgeoisie without a proletariat. The bourgeoisie naturally conceives the world in which it is supreme to be the best" (70–71).

As we will see, the transmission of bourgeois values was openly a goal of legislation prohibiting public violence to animals on the streets of urban Europe. Kindness to animals came to stand high in the index of civilization. It formed part of the project of civilization. The barbarian others—the urban working classes, continental peasants, southern Europeans, Irish Catholics, Russians, Asians, and Turks—were defined in part by their brutality to beasts.

Observers noted that animals were protected by law in England before slavery was abolished and before children were protected from the worst exploitations of the factory system. The RSPCA was accused of humanitarian inconsistency. It is true that only in 1833 were children under nine prohibited from working in factories and the work hours of older children regulated. Although the slave trade was abolished in 1807, slavery itself was legal throughout the Empire until 1833. It is clear, however, that the protection of animals against public cruelty was part of an expansive process of reform. Martin's Act of 1822 and the more inclusive animal protection act of 1835, which included dogs and cats—like the temperance movement, the ragged school movement, and the first suffrage reform act of 1833—were responses to the advance of capitalism. In a more general way they were a part of that modernization of state and society which characterizes English culture in the first half of the nineteenth century.

As James Turner suggests (1980), however, it would be a mistake to see the origins of the animal protection movement in industrialization per se. Not only, as we have seen, did the movement to protect animals from cruelty begin in the seventeenth century, but industrialization itself did not distance the English from animals. Ponies were used in mines, horses along canals and for the building of railroads. Horses provided transportation in cities for most of the century, as did dogs, which pulled carts until 1839 in London and until 1854 elsewhere. The cavalry remained an important part of armies until World War I. Veterinary schools were founded to train students to treat horses and livestock.

The animal protection movement in the nineteenth century was only indirectly related to a romantic view of nature, which pitted it against urban industrialization. It had an obvious though not exclusive class dimension and forms a chapter in the history of attitudes toward violence. A London cabdriver's out-burst of anger, for instance, that resulted in his beating to death an old, weak horse on a London street is a recurring image of animal protection literature. From the point of view of the RSPCA and its sympathizers, the killing was a dangerously irrational act. Beating a dying horse will not make it work—those who are vicious to animals will be murderous to others. From the point of view of workers and their advocates, however, the attempt to get a cab moving again is desperately rational, as Anna Sewell made clear to contemporaries in *Black Beauty*. Fares were needed for survival.

The RSPCA attacked the recreations as well as the livelihoods of the London poor. Dogfights as well as dog-carts were denounced, but foxhunting by the professional and landed classes was left alone. Violence was to be sequestered, hidden away from the view of those susceptible to its pernicious influence. This explains the attempt in the first part of the nineteenth century to move London slaughterhouses to the periphery of the city, so the sights and sounds of dying animals would not disturb neighborhood life.

Two principles informed the animal protection movement in the nineteenth century. The first was familiar to seventeenth- and eighteenth-century reformers: We have a duty to God to treat well each of his creatures who are dependent upon us. People should not cause animals unnecessary pain. The second was the need to quarantine violence, because like disease it "communicates an immoral contagion of the worst and most virulent kind among those who witness it" (Harrison 120). As in the seventeenth century, the protection of animals in the nineteenth century figured in a formula of social control.

By the last third of the century, the conflicts which animal protectionism defined were changing. Through the issue of vivisection, a challenge to scientists and materialism was formulated which anticipated twentieth-century concerns. In addition, anti-vivisection became an element in the feminist movement in Britain and France especially. As we will see, a strong conservative element appeared in this new episode in the history of animal protection, in Germany and Switzerland most dramatically as the anti-vivisection movement became linked with antisemitism.

Vivisection, experimentation on live animals to understand the mechanisms of the liver, the pancreas, the spleen, and other organs, was developed particularly by French and German physiologists. One of the most important in France was Claude Bernard, whose *Introduction to the Study of Experimental Medicine* was widely influential. Vivisectionists operated mainly on small animals, though sometimes horses were used in veterinary schools. Because of the availability and size of dogs, they were favored animals of vivisectionists. The image of the faithful and loving family dog begging for his life in the laboratory of the vivisectionist was also favored in anti-vivisectionist propaganda. The idea that the family pet, when lost, would end up on the vivisection table frightened children well into the twentieth century and linked this aspect of animal protection with the newly widespread practice of petkeeping in nineteenth-century Europe.

The anti-vivisection movement was important in western Europe from the 1870s to the beginning of World War I. It was first of all an expression of conflict within the elite over the purpose of science and the possibilities of its regulation. The question of whether scientists should be regulated was debated in Britain, France, and Germany. In Bismarckian Germany, anti-vivisectionists spoke from the conservative and center opposition, repeatedly petitioning the Reichstag in the 1880s and 1890s to abolish vivisection, but to no avail. The practice was left to the discretion of German scientists until the Nazi takeover of the state. In Britain, the Act to Amend the Law Relating to Cruelty to Animals in 1876 was the world's first restriction of vivisection. It established a licensing requirement. Hostile public opinion kept open the debate on vivisection, however. Both sides maintained a very active propaganda war until 1913, when the Royal Commission on Vivisection's Final Report upheld the practice of vivisection but subjected it to continued legal control. In France, the question of whether to restrict vivisection was studied by the Academy of Medicine and by a committee of the Société protectrice des animaux. As in Germany, and unlike in Britain, vivisection remained self-regulated in France in the nineteenth century.

Within the established animal protection societies in England and France, a consensus formed that vivisection could be allowed if animals were caused no unnecessary pain. The use of anesthesia was urged. French science was dependent on vivisection to an extent the British refused, however. Protests against the visits of French physiologists to Britain became debates over the costs of modernity, with British public opinion once again granting the English superiority over the French by virtue of their greater kindness to animals.

Vivisection stimulated an examination of the relationship between scientists and the state. More dramatically, it raised questions about women's roles and about the meaning of being female. Anti-vivisection is linked, therefore, to the development of feminism in the late nineteenth century. Some historians suggest that the anti-vivisection movement empowered women by providing them with leadership positions in volunteer organizations and a voice in the public sphere. Within the RSPCA and the Société protectrice, women played a largely decorative or behind-the-scenes role. But the leadership of anti-vivisection societies included very effective women. The Victoria Street Society for the Protection of Animals from Vivisection (established in 1876) was led by Frances Power Cobbe—famous already for her propaganda war in Florence against the German physiologist Moritz Schiff. Marie Huot and Maria Deraismes in France led the Ligue populaire contre les abus de la vivisection (Popular league against the abuses of vivisection). Within the Parisian animal protection society the issue of vivisection moved ordinarily demure female members to speak out in opposition. Marie-Espérance von Schwartz was an ally of Ernst von Weber, who founded the Internationale Gesellschaft zur Bekämpfung der wissenschaftlichen Thierfolter (International society for combat against scientific torture of animals) in 1879, and a member of the society's directing committee (see Elston; Tröhler and Maehle; and Kete 5–21).

Mary Ann Elston points to the influence of women within the RSPCA (262–63). By establishing animal refuges, they saved dogs from hard-hearted workers in mid century and from evil scientists in the last part of the century. Of course, men were leaders in the anti-vivisection movement, too. Its strongest supporters in England included men on both sides of the question of woman's suffrage. In Germany, its most famous supporter may have been Richard Wagner, who, as Tröhler and Maehle note, famously claimed not to want to live in a world "in which 'no dog would wish to live any longer'" (176).

Some women claimed an identification with animals mistreated by scientists, an identification which galvanized feminist consciousness. Women, like animals, were at the mercy of male rationalism. As Coral Lansbury asserts in *The Old Brown Dog*, Claude Bernard himself had "described nature as a woman who must be forced to unveil herself when she is attacked by the experimenter, who must be put to the question and subdued" (162–63). In anti-vivisection imagery, as well, the vivisector appears as a sexual predator, sadistically enjoying a perverse pleasure in causing prostrate animals pain. This is the image which appears in *Gemma, or Virtue and Vice* by Marie-Espérance von Schwartz, in *The Beth Book* by Sarah Grand, and in other works which Coral Lansbury compares with pornography (12–29).

The anti-vivisection movement emphasized the importance of feeling as a guide to understanding, rather than the use of the scientific method. It thus could serve as an interrogation of materialism, a rethinking of the aims and means of science. But the identification of women with animals abused by male science drew upon essentialist notions of female identity. It spoke to conventional binaries—woman and nature, man and culture, feminine emotion and masculine reason—and, to an important degree, served a conservative purpose. The anti-vivisection movement included suffragists in England, but also anti-suffragists, and stood with conservatives in Bismarckian Germany.

To those who promoted the anti-vivisection movement, society could seem divided into ruthless men of science and women, whose maternal roles of childbearing and nurturing gave them a special affinity with the world of nature and allowed them to critique the experimental method. In Germany, especially, this critique of materialism came to focus on Jews. In the minds of German and Swiss anti-vivisectionists, it was Jewish doctors who practiced vivisection and "Jewish" attitudes toward animals which allowed for it. Schopenhauer had argued earlier in the century that, as Tröhler and Maehle put it, "it was time that the 'Jewish' view regarding animals came to an end" (151). For antisemites like Wagner, this "Jewish" attitude was expressed in both vivisection and kosher butchering. (Its reverse, vegetarianism, was strongly promoted in Bayreuth.) The journal of the German anti-vivisection movement, *Thier- und Menschenfreund*, as Tröhler and Maehle note, strongly supported the abolition of kosher butchering, which was achieved in Switzerland in 1893 and by the Nazis in April of 1933 (176). The image of the kosher butcher practicing a private, bloody orgiastic rite was much like the image of the vivisector, as a viewing of the Nazi propaganda film *The Eternal Jew* makes clear.

Keith Thomas speaks in *Man and the Natural World* of the dethronement of humans, a process, he claims, which begins in early modern Europe and continues through the nineteenth century (165). In the twentieth century the abandonment of the principle of the sanctity of human life and the hierarchy it presumes led both the radical right and the radical left to rethink the relationship between humans and animals.

In "Understanding Nazi Animal Protection and the Holocaust," Arnold Arluke and Boria Sax discuss Nazi animal protection legislation in the context of the Nazi revolution of state and society. One of the first laws passed by the Nazis in April 1933 prohibited kosher butchering. Soon afterward, vivisection was first abolished, then restored with restrictions. Nazi animal protection extended far beyond these two overtly antisemitic acts, however. Laws covered the treatment of lobster and shellfish by cooks. To reduce their suffering, lobsters were to be thrown only one by one into rapidly boiling water. Another provision protected horses that were being shoed. Endangered species such as the bear, the bison, and the wild horse were protected.

Nazi animal protection legislation was not much more comprehensive than the British, Arluke and Sax point out, but, clearly, the Nazi understanding of the relationship between humans and animals was profoundly distinct from traditional European beliefs. Nazism "obliterated" "moral distinctions between animals and people," Arluke and Sax explain, allowing a reordering of the chain of being (23). Some animal species rested above some human "races." So Aryans, German shepherds ("deliberately bred to represent and embody the spirit of National Socialism" [14]), beasts of prey, and Teutonic acorn-eating pigs were far superior to subhuman "races." Jews were vermin that needed to be killed, as six million were in the death camps and the ravaged villages of eastern Europe.

The Nazi understanding of the natural world stands in contrast to the traditional and modern European understanding of the human/animal divide, which in privileging humans enabled a civilizing mission to be undertaken on behalf of the brutal but still human "other," the urban working classes, the primitive Mediterranean peoples, and the peasant. The Nazis' elision of the human/animal divide allowed them to murderously express the superiority of some humans at the expense of others. (The Soviets, on the other hand, maintained Marx's nineteenth-century understanding of humans as distinct from other animals. Their destruction of the environment of large parts of eastern Europe, made apparent after the fall of Communism, speaks against the celebration of this view, as well.)

The animal liberation movement of the 1970s renewed debate about the social meaning of human relationships with animals. Peter Singer's *Animal Liberation* compared speciesism (a neologism) to racism and sexism. In each case—racism, sexism, speciesism—he argued, arbitrary characteristics are the signal for discrimination. In the case of the human species, our ability to reason is the excuse to oppress other species.

Singer's argument begins with a misreading of Mary Wollstonecraft. In the famous first paragraph of *Animal Liberation* he explains that "[t]he idea of 'the

Rights of Animals' actually was once used to parody the case for women's rights":

> When Mary Wollstonecraft, a forerunner of today's feminists, published her *Vindication of the Rights of Women* in 1792, her views were widely regarded as absurd, and before long an anonymous publication appeared entitled *A Vindication of the Rights of Brutes*. The author of this satirical work (now known to have been Thomas Taylor, a distinguished Cambridge philosopher) tried to refute Mary Wollstonecraft's arguments by showing that they could be carried one stage further. If the argument for equality was sound when applied to women, why should it not be applied to dogs, cats, and horses? The reasoning seemed to hold for these "brutes" too; yet to hold that brutes had rights was manifestly absurd. Therefore the reasoning by which this conclusion had been reached must be unsound, and if unsound when applied to brutes, it must also be unsound when applied to women, since the very same arguments had been used in each case. (1)

Wollstonecraft's point—shared by Condorcet and others—was that differences between men and women were only apparent, the result of different upbringings. It was society's expectations for men and women that led them to separate spheres. Taylor's "joke" lies in his intended readers' "recognition" that women are irrational, different from men, and as incapable as animals of participating in the public sphere of government and business. Singer's ploy is to associate himself with Wollstonecraft—who believed in the equality of men and women based on the universality of human nature—while, like Taylor, to forward an argument based on the premise of difference.

"Abilities and capabilities" cannot stand as the basis for equality, since "we can have no absolute guarantee that these capabilities and abilities really are distributed evenly, without regard to race or sex, among human beings" (Singer 4). For Singer equality is prescriptive, not actual, and capable of being extended to every living creature. Moral equality demands not the equal treatment of humans but that (following Bentham) "the interests of every being affected by an action are to be taken into account and given the same weight as the like interests of any other being" (5).

Thomas Jefferson's doubts about the reasoning ability of "Negroes," which he expressed in 1809 in a letter to Henri Grégoire, are brought to bear on Singer's argument. Jefferson wrote,

> Be assured that no person living wishes more sincerely than I do, to see a complete refutation of the doubts I myself have entertained and expressed on the grade of understanding allotted to them by nature, and to find that they are on a par with ourselves . . . but whatever be their degree of talent it is no measure of their rights. Because Sir Isaac Newton was superior to others in understanding, he was not therefore lord of the property or persons of others. (qtd. in Singer 6)

Singer asks us to see that "Thomas Jefferson, who was responsible for writing the principle of the equality of men into the Declaration of Independence," agreed with him that "the basic element—the taking into account of the interests of the being, whatever those interests may be—must, according to the principle of equality, be extended to all beings, black or white, masculine or feminine, human or nonhuman" (5).

Singer's presentation of his argument leads us to associate the animal liberation movement with the Enlightenment project of human rights. For Wollstonecraft and Jefferson, human nature was universal and distinctive, conditioned by the environment, and tending toward progress (see Matthews 53). But Singer reshapes this material to make an unexpectedly conservative point—that the condition of beings is both static and particular. The principle of equality demands equal consideration for interests based on existing "characteristics," Singer tells us. He continues, "[C]oncern for the well-being of children growing up in America [but not in Africa?] would require that we teach them to read; concern for the well-being of pigs may require no more than that we leave them with other pigs in a place where there is adequate food and room to run freely" (5).

Singer's logic fragments the line traditionally drawn between human and other beings by forwarding an argument of difference (potential or actualized) between individuals, between races, and between the sexes. His dismantling of Enlightenment universalist claims masks an anti-democratic stance (see Ferry's discussion of "deep ecology," 59–70). It leads us to conclude that we must, ethically, make decisions based on the "[c]haracteristics of those affected by what we do" (5). It allows one to privilege the non-human animal over the human, in a disquieting echo of Nazi conclusions: "This is why when we consider members of our own species who lack the characteristics of normal humans we can no longer say that their lives are always to be preferred to those of other animals" (21), Singer offers. His contributions to the field of medical ethics are thus both brave and disturbing.

In the animal liberation movement of the 1970s and 1980s in Europe, which Singer's book, along with Regan's *The Case for Animal Rights*, helped to promote, anti-vivisection again became a cause. Protestors investigated animal research at university and private laboratories and succeeded in bringing important cases of animal abuse to light. Older causes, such as the transportation of animals to slaughter, were taken up in England by the Compassion in World Farming (CIWF) group.

But older themes as well as older issues have prevailed in the late-twentieth-century animal protection movement and invite us to consider the extent to which the representations of animals constructed in the social conflicts of modernizing Europe still shape social conflicts today. Hilda Kean notes in *Animal Rights* that in the CIWF campaign against Parisian Muslims' slaughter of sheep for the festival of Eid el Kebir, the British provenance of the sheep figured strongly (21). Kean notes, too, that recent campaigns against vivisection in England highlighted the fact that the animals used in British laboratories were imported from southern Europe, southeast Asia, and the Caribbean, speaking to an earlier British sense of being uniquely civilized in the care of nature. In England, as well, the fight to abolish foxhunting seems likely to continue along class and rural/urban lines.

It seems clear from other late-twentieth-century events, such as the outbreak of mad cow disease and the ensuing British-French enmity, that Europeans will continue to find meaning in their relationship to animals along the lines of earlier structures of thought. Regional enmities as well as a sense of human guardianship of nature still prevail. Whether the logic of "dethronement" will also have social consequences in the twenty-first century, as it most dramatically did in Nazi Germany, is more difficult to know. What seems certain is the potential of thinking about animals to construct scripts of oppression as well as liberation, as the historiography of attitudes toward animals in modernizing Europe bears out.

Bibliography

Agulhon, Maurice. "Le sang des bêtes: Le probleme de la protection des animaux en France au XIXème siècle." *Romanticism: Revue du dix-neuvieme siècle* 31 (1981): 81–109.

Arluke, Arnold, and Boria Sax. "Understanding Nazi Animal Protection and the Holocaust." *Anthrozoös* 5.1 (1992): 6–31.

Blackbourn, David. *The Long Nineteenth Century: A History of Germany, 1780–1918.* New York: Oxford University Press, 1998.

Bruford, W. H. *Germany in the Eighteenth Century: The Social Background of Literary Revival.* Cambridge: Cambridge University Press, 1971.

Cowan, Alexander. *Urban Europe, 1500–1700.* London: Arnold, 1998.

Darnton, Robert. *The Great Cat Massacre and Other Episodes in French Cultural History.* 1984. New York: Vintage, 1985.

Delort, Robert. *Les animaux ont une histoire.* Paris: Seuil, 1984.

Elston, Mary Ann. "Women and Anti-vivisection in Victorian England, 1870–1900." Rupke, *Vivisection,* 259–94.

Ferry, Luc. "Think Like a Mountain." *The New Ecological Order.* 1992. Trans. Carol Volk. Chicago: University of Chicago Press, 1995, 59–70.

French, Richard D. *Antivivisection and Medical Science in Victorian Society.* Princeton: Princeton University Press, 1975.

Fudge, Erica. *Perceiving Animals: Humans and Beasts in Early Modern English Culture.* New York: St. Martin's, 2000.

Harrison, Brian. *Peaceable Kingdom: Stability and Change in Modern Britain.* Oxford: Oxford University Press, 1982.

Hay, Douglas. "Poaching and the Game Laws on Cannock Chase." *Albion's Fatal Tree: Crime and Society in Eighteenth-Century England.* Ed. Douglas Hay et al. New York: Pantheon, 1975. 189–253.

Hearne, Vicki. *Adam's Task: Calling Animals by Name.* New York: Knopf, 1986.

Kean, Hilda. *Animal Rights: Political and Social Change in Britain since 1800.* London: Reaktion, 1998.

Kete, Kathleen. *The Beast in the Boudoir: Petkeeping in Nineteenth-Century Paris.* Berkeley and Los Angeles: University of California Press, 1994.

Lansbury, Coral. *The Old Brown Dog: Women, Workers, and Vivisection in Edwardian England.* Madison: University of Wisconsin Press, 1985.

Malcolmson, Robert. *Popular Recreations in English Society, 1700–1850*. Cambridge: Cambridge University Press, 1973.

Marx, Karl, and Frederick Engels. *The Communist Manifesto: A Modern Edition*. Intro. Eric Hobsbawm. London: Verso, 1998.

Matthews, Richard K. *The Radical Politics of Thomas Jefferson: A Revisionist View*. Lawrence: University Press of Kansas, 1984.

Munsche, P. B. *Gentlemen and Poachers: The English Game Laws, 1671–1831*. Cambridge: Cambridge University Press, 1981.

Phineas, Charles [pseud.]. "Household Pets and Urban Alienation." *Journal of Social History* 7.3 (1974): 338–43.

"Protection des animaux: Encouragements, demandes par la Société protectrice des animaux, diffusion de la loi Grammont," circulaire du 10 mars 1894. 1863–1900. Archives Nationale, Série F17, 11696–97.

Regan, Tom. *The Case for Animal Rights*. Berkeley and Los Angeles: University of California Press, 1983.

Ritvo, Harriet. *The Animal Estate: The English and Other Creatures in the Victorian Age*. Cambridge, Mass.: Harvard University Press, 1987.

——. *The Platypus and the Mermaid and Other Figments of the Classifying Imagination*. Cambridge, Mass.: Harvard University Press, 1997.

Rupke, Nicolaas A., ed. *Vivisection in Historical Perspective*. New York: Croom Helm, 1987.

Schama, Simon. *Citizens: A Chronicle of the French Revolution*. New York: Knopf, 1989.

Singer, Peter. *Animal Liberation*. 1975. 2nd ed. New York: New York Review Books, 1990.

Thomas, Keith. *Man and the Natural World: A History of the Modern Sensibility*. New York: Pantheon, 1983.

Tocqueville, Alexis de. *The Old Regime and the French Revolution*. Trans. Stuart Gilbert. Garden City, N.J.: Doubleday, 1955.

Tröhler, Ulrich, and Andreas-Holger Maehle. "Anti-Vivisection in Nineteenth-Century Germany and Switzerland: Motives and Methods." Rupke, *Vivisection*, 149–87.

Tuan, Yi-fu. *Dominance and Affection: The Making of Pets*. New Haven: Yale University Press, 1984.

Turner, James. *Reckoning with the Beast: Animals, Pain, and Humanity in the Victorian Mind*. Baltimore: Johns Hopkins University Press, 1980.

Weber, Eugen. *Peasants into Frenchmen: The Modernization of Rural France, 1870–1914*. Stanford: Stanford University Press, 1976.

Woloch, Isser. *Eighteenth-Century Europe: Tradition and Progress, 1715–1789*. New York: Norton, 1982.

3   Dog Years, Human Fears

Teresa Mangum

The end of the nineteenth century in Britain marked the beginning of a pre-occupation with old age that proliferates in Western culture today. By the 1870s, forces had gathered that continue to shape more recent conceptions of aging. The British government debated the duty of the nation to create pension plans for the elderly; the brisk industrial economy prompted mandatory retirement policies. Sociological surveys revealed that a shocking number of older Britons were living in ill health and dire poverty throughout the country even as geriatrics emerged as a distinct medical field concerned with the diseases of old age. Moreover, a very elderly Queen Victoria continued to rule the British Empire, a point visibly impressed upon her public during the two Jubilees celebrating the fiftieth and sixtieth years of her reign. Among the myriad cultural responses to a growing awareness of the nature and conditions of late life, none is more intriguing than the emergence of a surprising new literary voice: that of the aged autobiographical dog.

Why did so many nineteenth-century readers willingly accept the fiction of the speaking animal? In the literary milieu of fictive animals like Black Beauty and Beautiful Joe, animals became impossibly positioned as fully articulate subjects with a great deal to say to their human readers and listeners. More than any others, it was the canine point of view that took center stage in a host of British Victorian novels. Self-proclaimed autobiographies of show dogs, hunting dogs, lost or escaped dogs, and neglected, even tortured dogs took the form of novels, of short stories published in periodicals and collected in anthologies, and of poetry written about and to dogs. Ignoring both the charm of puppies and the prowess of midlife dogs, Victorian writers of dog autobiographies focused for the most part on the old dog—short-lived by human chronology but old in "dog years," that fiction by which humans calculate the lives of dogs in an acceleration of human time. The pressure of this condensation only intensified the poignancy of the dog as it came to stand in for those qualities people sought in fellow humans—attentiveness, unconditional love, courage, loyalty so unwavering it persists even in the face of death. In effect, imagined emotional "doggedness" ensured fictional subjectivity. Not surprisingly, in a culture receptive to dog narrators, the loss of a dog often provoked an answering voice. Magazines of the period regularly published elegies and epitaphs in which humans attempted to speak to and for deceased canine companions.[1] As Kathleen Kete points out in her fascinating *The Beast in the Boudoir: Petkeeping in Nineteenth-*

Century Paris, pet cemeteries were established in Paris (as they were in London), where human responses to pets could be literally carved in stone. In this essay, I discuss how dogs in Victorian Britain came to be saturated in subjectivity and why that subjectivity was so often marked by associations with old age and death, a connection so powerful that "dog years" has become shorthand for a particular compression of time based on the rapidity—in human terms—with which dogs grow old and die.

The history of the dog in nineteenth-century Britain is itself a tale of identity formation. Each repositioning of the dog in its relations to humans provoked the desire to project a readable subjectivity onto the animal, and this desire was at least as strong as any impulse to comprehend whatever nature inheres in dogs. Harriet Ritvo provides a wealth of examples in *The Animal Estate* of the contexts into which dogs were placed in the period (82–121). Extrapolating from Ritvo's examples, we can see how particular contexts then function as discursive landscapes which give new meaning to aspects of dog behavior from the perspective of Victorian pet owners and observers. For instance, Ritvo shows that in the second half of the century, exhibitions of dogs became enormously popular, leading to the formation of the British Kennel Club in 1873 and an obsession with establishing clearly distinguished breeds and sub-breeds, of fox terriers and Skye terriers rather than just terriers, for instance. Breeding led to a campaign for what Stephen Budiansky recently referred to as dog "eugenics," upsetting recent readers of the *Atlantic Monthly.* Heretofore hunting and working dogs had served humans (a cart-pulling dog bleakly figures in *A Dog of Flanders: A Christmas Story* [1891] a lachrymose children's novel by the popular author Marie de la Ramée, who published as Ouida). Very differently, kennel club shows positioned the dog as a spectacle, even a performer. Moreover, even as the wealthier classes took up the breeding and training of show dogs, the early-nineteenth-century general public delighted in the spectacle of "Carlo, the Performing Dog" on the London stage, an act one can still observe today as one of the featured displays of the London Theatre Museum. In his daily performances, Carlo leapt from a high plank to drag a drowning young girl from a tub of water, a feat he describes in his alleged autobiography. In the shift from display to theatrics we find an analogue to the shift from fiction about dogs to fiction in the voice of dogs.

One could argue for similar slippage when dog owners moved beyond brass collars and jeweled leashes to clothing, and dogs entered the arena of fashion. In an 1896 article titled "Dandy Dogs," William G. Fitzgerald of the *Strand Magazine* describes the Dogs' Toilet Club of New Bond Street. The dogs' wardrobes include morning dress, riding costumes, and evening gowns; Kete describes similar excesses in her account of Parisian dogs (81–86). According to Fitzgerald, dogs of fashion not only had their own day spas and wardrobes, but their own society veterinarians. One of the most in demand, a "Mr. C. Rotherham," received unique dog narratives in which dog owners, usually women, inhabit the voices of their dogs, to sidestep responsibility for their pets' ill health. Fitzgerald quotes one such dog narrator:

—Belgrave Square, W.,
        22nd January, 1896

DEAR DR. ROTHERHAM,—As they say in American, I feel "real sick" this morning; so mother tells me to write and ask you to call here as early as possible after receiving this. . . . I will tell you how I feel, so that you may in some measure be guided in your treatment of my indisposition. You must promise not to tell mother, but she gave a dinner last evening, and I *did* enjoy myself. I had *such* a lot of nice things! Do you think it is possible for them to have made me feel as I do? I was in great pain during the night, so that poor mother and myself did not have a wink of sleep." [AND SO FORTH]

Your grateful patient,
NIGGY (547)

Even earlier, as documented by Brian Harrison among others, reformers had begun to include animal abuse and neglect among their other concerns, leading to the formation of the Society for the Prevention of Cruelty to Animals in 1824 by Richard Martin and William Wilberforce, leaders of the abolitionist movement. In the 1870s a children's auxiliary called Bands of Mercy (now Animal Defender) was organized along the same lines as the children's temperance group, the Bands of Hope. (One of the most popular songs sung by these children was "Only a Cur," which also forms the title of the opening chapter of the dog "autobiography" *Beautiful Joe*, written by Marshall Saunders [1893].) Similarly, in the second half of the century, as Coral Lansbury documents in *The Old Brown Dog: Women, Workers, and Vivisection in Edwardian England,* a host of anti-vivisection organizations, anti-muzzling organizations, and animal "homes" (where lost and abandoned animals were cared for or "cured" of care through euthanasia) were established. As Harrison notes, these groups produced endless statistical reports and published anecdotes of cruelty to animals in a century-long effort to convince members of Victorian society and Victorian lawmakers that at the very least cruelty to animals was inhuman (110–22). More radically, these examples were also used to argue that animals were subjects in their own right, capable of not only pain but also fear, misery, memory, and mourning—of grieving both for their own circumstances and for the loss of their human companions. This second imperative also led to some of the most powerful narratives "authored" by animals. Anna Sewell's *Black Beauty* (1877), which features a horse rather than a dog, is perhaps the best known of these protest novels. It was soon followed by *Beautiful Joe,* written by an American but, like *Black Beauty,* popular in both Britain and the United States.

Animal subjectivity is articulated with special force in visual representations of dogs. Like dog autobiographies, the emergence of the dog portrait signals the transformation of the domestic animal into the animal companion, even the animal family member. These categories follow from evolving human-animal relations which gradually position dogs at the center of rather than as marginal or adjacent to domestic life. In his *Dog Painting, 1840–1940: A Social History of the Dog in Art,* William Secord discusses the paintings of Edwin Landseer,

Richard Ansdell, Gourlay Steel, George Earl, John Emms, John Noble, Frank Paton, Briton Riviere, and Arthur Wardle, among many others. Though the portraits include every conceivable kind of dog at every stage of a dog's existence, among the most popular pictures were those that featured dogs, presumably old themselves, in attitudes of mourning over the deaths of their human "masters," the term Victorian writers most commonly applied to dog owners. In fact, the dog's reputation for loyalty, fidelity, and undying faith recurs in image after image of the Victorian dog as not merely one mourner but chief mourner. In these instances, we can read the dog simultaneously as a melancholy stand-in for the aged human companion, now dead and presumably forgotten by less devoted human survivors, and as an increasingly loaded signifier of its own heretofore unarticulated self—a conscious canine capable of the intricate, textured memories, the deep melancholy, and the temporally variegated imagination that inspires lonely visions of a solitary future. The master of animal painting, Sir Edwin Landseer, produced two of the most popular versions of this dog narrative: *The Poor Dog* (or *The Shepherd's Grave*) in 1829 and *The Old Shepherd's Chief Mourner* in 1837. Probably the most famous dog painting of the Victorian era, the latter, affectionately known to the public as *Old Shep*, depicts the shepherd's desolate collie resting its head on the coffin of its beloved master, surrounded by the accouterments of their shared work life and lit by an open window through which the shepherd's spirit has presumably escaped. Briton Riviere depicts the same theme in *Requiescat*, exhibited at the Royal Academy in 1888. Here a hound sits at attention alongside a bier, covered by beautiful brocade, upon which his armored master lies in state. The dog's eyes are trained in anxious attention upon what he can see of the knight's upturned face as he waits devotedly for a return to duty. Of course, many of us have grown up with the RCA Victor mascot, based on an 1899 painting by Francis Barraud titled *His Master's Voice*. Barraud claimed that when he played his dead brother's voice to his brother's pet terrier, Nipper, the dog would strike a puzzled pose of longing before the gramophone. Ears half-cocked, face drawn down by sagging, wrinkled jowls, Nipper attempts to smell as well as hear his lost master. What twentieth-century viewers interpreted as a dog puzzled by the marvels of technology, Victorians perceived as a poignantly longing canine subject eager, perhaps, both for the voice of his master and for a voice in which to express inarticulate, unending faithfulness and grief. Both paintings, in fact, resemble yet another powerful and popular metonym for Victorian grief, a painting from mid century called *Old Faithful* executed by Charles Dickens's son-in-law Charles Perugini. Centered in the painting, though half-turned away from the viewer, an elderly woman in the sepia tones of long mourning sits quietly, sadly reflecting upon what must be her husband's grave. Is the old woman by analogy a dog, just as old mourning dogs assume humanity, albeit humanity with little gender or agency? We see a real-life version of this visual analogy linking old dogs to old humans and hence to old days in a glossy sepia postcard print of Edward VII from a photograph taken by Thomas Heinrich Voigt around 1900 and circulated after the king's death. Voigt's photograph shows the elderly King Edward sitting beside Caesar, his companion during the last seven years of his life. Caesar won the fame on which the postcard sales depended by following the king's coffin from Westminster Hall to Windsor; the faithful Caesar dominated coverage of the king's end.[2]

Victorian commentators could be very perceptive about the reasons why dogs so readily came to be associated with late life, the passage of time, and death itself. An anonymous writer for the *Saturday Review*, in an 1889 essay titled "A Drunken Dog," makes these connections fairly directly. The dog, he writes,

> has been on such extremely intimate terms with man, that through thousands of generations he has acquired an amount of humanity, which to a nice observer is very astonishing. The thousands of generations have rather a melancholy aspect. The dog is so very short-lived. He is aged at fifteen years, as old in point of decrepitude as a horse at thirty, more so than a man at eighty. It is sad to think for how short a time we have this prime favorite with us, and what lamentations are poured over his early grave. . . . Perhaps nature has designed him to wear himself out quickly, so that he shall not live long enough to know too much, to learn to speak, to write—in short, to rival her proud piece of work, man, as he might if he had fifty years instead of fifteen to do it in. He is an old decrepit person, with great experience, but with his faculties all used up, when man is just escaping from childhood. (703–704)

Even as the writer acknowledges the potential for humans to feel the loss a dog's life portends, he nervously resists the possibility of the dog becoming too human, able to "rival" humans, if only in age. The slip from young dog to "old decrepit person" is especially telling. Here "dog years" reassure (human) readers of their own tenacious hold on life even as animal entities merge with those of aged—hence soon-to-die—humans. The writer, in fact, suggests yet another reason why dogs prompt thoughts of old age. Even as old dogs are positioned as increasingly human in portraits and fiction, older humans were often described in terms suggesting a loss of humanity.

Literary sources provide an equally vivid context in which to study the emerging and often aged subjectivity of the dog, and again the Victorians themselves turned to these sources in attempting to understand their relations to dogs (see "Animals in Novels"; and "Dogs of Literature"). Literary representations of dogged loyalty in old age and to human old age reach as far back as the *Odyssey* and were much-loved plot devices of Sir Walter Scott. In *Old Mortality,* for example, when the wandering elderly hero returns home, only his faithful dog recognizes him. Scott returned the favor in his lifetime by memorializing his long-time canine companions. One biographer recounts his doglike loyalty to his hound, Maida:

> While the dog was still alive, though failing, and only now and then raising a majestic bark from behind the house at Abbotsford, a statue of him was erected at the door. Those were the days when Scott used to stroll out in the morning to visit his "aged friend," who would "drag his gaunt limbs forward painfully, yet with some

remains of dignity, to meet the hand and loving tone of his master," as he con-
doled with him on being "so frail." But the end came at last, and Maida died
quietly one evening in his straw bed, of sheer old age and natural decay. (qtd. in
"Dogs of Literature" 480–81)

At the same time, Scott seems more alert than most writers to the empty comfort of "dog years." Another biographer quotes Scott's musings on dog years and
their implications for humans: " 'The misery of keeping a dog,' says he, 'is his
dying so soon; but, to be sure, if he lived for fifty years, and then died—what
would become of me?' " ("Dogs of Literature" 479). A curious reversal of this
idea, but similarly suggesting the odd interchangeability among old dogs, old or
lost human associates, and old time, is also associated with another famous
author. In an article titled "Relics of Emily Brontë," published in *The Bookman*
in 1897, Clement Shorter includes Brontë's sketches and watercolors of her various dogs among the relics.

However it is the animal memoir in which the old dog as narrator takes pre-
eminence. In Britain, Gordon Stables, a surgeon in the Royal Navy, wrote a number of books about animals, including his most successful, *Sable and White: The
Autobiography of a Show Dog* (1894). *Diomed: The Life, Travels, and Observations of a Dog* (which I have not yet located) was published by John Sargeant
Wise in 1897 and mistakenly described by a reviewer in *The Bookman* (Septem-
ber 1897) as the first "literature from the point of view of animals." The reviewer notes, "The narrative is told in retrospect, and once in a while a rheu-
matic twinge arrests the old dog's thoughts, causing him to reflect sadly on the
vanities of life" (Tracy 71–72). Beginning in the 1890s, Rudyard Kipling wrote
several short stories narrated by elderly dogs in abysmal dog English nearly too
embarrassing to quote. (His dogs are admirably if absurdly doggish, unlike most
dog narrators, fawning on their "Own Gods" and focusing on food and other
animal pleasures.) Numerous short stories published in nineteenth-century periodicals feature aging animals which embody fused fantasies of home, nation-
hood, memory, and nostalgia, on one hand, and chastisement of human greed,
insensitivity, cruelty, and commodification of animal life, on the other. Perhaps
even more telling, a *Punch* series with titles such as "A Dog on His Day" (1894)
and "Page from a Dog's Diary" (1899) suggests that by 1899 dog narrators were
pervasive enough to merit mockery.

Inspired by the success in 1877 of *Black Beauty*, the "autobiography" of
an elderly horse, the American Marshall Saunders wrote *Beautiful Joe: An Autobiography*. As Joe, the canine narrator, explains to the reader in the opening
pages of the novel,

> I am an old dog now, and writing, or rather getting a friend to write, the story
> of my life. I have seen my mistress laughing and crying over a little book that she
> says is a story of a horse's life, and sometimes she puts the book down close to my
> nose to let me see the pictures.
>
> I love my dear mistress; I can say no more than that; I love her better than
> any one else in the world; and I think it will please her if I write the story of a

dog's life. She loves dumb animals, and it always grieves her to see them treated cruelly. (14)

In the novel, the "dumb" voice of old age offers a means of asserting canine modesty and cushions the documentation of human cruelty. Old age also justifies the prosy sermons the narrator often preaches. Joe's doglike gratitude for the least human kindness, his self-denigrating tendencies, his dependence, even his tendency toward abjection are expressed as characteristics due simultaneously to his being an animal and his being old, as the two aspects of identity merge into an extended plea for readers' compassion for helpless beings.

The narrator of Gordon Stables's *Sable and White*, another self-styled "autobiography" of a dog, engages with old age in less predictable ways. Named for the heroic dog of epics, the Scottish collie Luath spins a picaresque tale of early abuse, terrifying train rides, happy homes, dog exhibitions (and the ways in which dog shows encourage the abuse of animals), being stolen, escaping a vivisectionist, a rescue by a showman who teaches Luath to "act," and eventual reunion with a beloved family. Illustrations picture not only the elegant, elderly collie narrating the story but his dog auditors. Here humans are relegated to the status of eavesdroppers on an imagined canine world of conversation and reminiscences. Time, aging, and necessary death—preferably, it seems, through euthanasia—are recurring subjects of dog discourse. Luath notes, for example, " 'What can dogs know about death?' some humans ask. A deal more than such humans imagine," continuing "As to age, Chummie, that is relative. A dog's years are shorter. A dog, two years old, is of the same age as a man of twenty. We learn more quickly the little we may know. . . . " But, Luath adds, almost contemptuously, that what he calls "the microbe man" "may be all at home in politics or algebra, but in a stubble field among the partridges, which is the nobler animal, that blind pottering old biped, or the noble Irish setter?" (70). This separate-but-equal principle obtains throughout the novel in quick, gentle vignettes of animals and humans quietly negotiating with old age together. The novel thus often intertwines the lives of old people and old dogs, suggesting an alternative narrative to those paintings in which dogs serve as mourners rather than companions in the last stages of life—their own lives and the humans with whom they live.

This same narrative of companions in aging forms a comic plot in the narrative poem "My Old Dog and I," from *Blackwood's Edinburgh Magazine* (1833). In the poem, the human narrator imagines herself in dialogue with her dog. The dog has not yet progressed to the point of carrying its own narrative, so the poem frames the voice of the dog with the voice of the human narrator—a more obvious acknowledgment of the ventriloquism to which all dog narrators are of course subject. The dog uses its age to con the human speaker into attending to its wishes for a walk. When she protests that he used to walk on his own, he "answers":

> Besides—I hate to walk alone —
> My eyes grow very dim;

I'm hard of hearing, too—a fly
    Might knock me down, so weak am I
        In ev'ry trembling limb.
And now, vile curs make sport of me —
        Vile creatures—but last week
Pounced on my back an old fat hen,
And peck'd me, till I howl'd again
        At every spiteful tweak. (lines 31–40)

Though the narrative ends with the dog's witty manipulation of its "master," the poem first detours into an acknowledgment of their mutual experience of late life:

Bear with me, Mistress!—I was not
        Always so curst a creature —
Perhaps old age, that on me gains
So fast, with all its aches and pains,
        Has something changed my nature,
But not my heart. I've served you now
        These eighteen years, well nigh —
Borne all your humours—(for *you*, too,
Mine honour'd Mistress! have a few,) —
        You'll own right lovingly;
Shared all your good and evil days —
        (Much evil have we known!)
Loved those you loved, and mourn'd them too,
And miss'd them long, as well as you;
        And now we're left alone,
I do my best, my very best,
        To please and cheer you still;
Though weak and weaker ev'ry hour
Becomes your poor old servant's power
        To prove his loving will. (lines 88–110)

The poem is poignant, yet witty too. Appearing much earlier than many of the paintings and texts I have discussed, the poem picks up on an alternative set of assumptions about dogs—they represent a kind of fawning, flattering cunning. Moreover, the poem posits a problem for dog subjectivity that fiction resolutely avoids. In becoming more human, dogs promise to become, well, more human—demanding, capable of imposing guilt, disagreeable, needy, inconvenient—and able to put those human qualities into words along with their more canine considerations.

To turn back to my opening question, why then is the dog given voice in such human fashion in the nineteenth century? Histories of the Victorian novel conventionally argue that the period marks the high point of philosophic and aesthetic realism in that many novels turn their attention to the conflicts of the middle and working classes; social and domestic problems predominate; the laws of science (or domestic management) structure metaphors as well as many fictive relationships; and emotions are stirred to serve the interests of sense rather than sentiment. I would argue, however, that the popularity of canine narrators marks the meeting of several seemingly inimical impulses that engaged nineteenth-century readers. First, dog narrators look backward to the late-eighteenth-century sentimental novel, a form which remained popular through the nineteenth century despite being criticized as a debased, emotionally exploitive, feminized form which sought to awaken sensibility and benevolence rather than reason.[3] In effect, the old dog is old in part because so many authors use the figure of the speaking dog to invoke past tastes in order to perform cultural work on behalf of animals in the Victorian present. As Janet Todd explains in her study of eighteenth-century fiction, sentimental novels were based on

> a belief in the appealing and aesthetic quality of virtue, displayed in a naughty world through a vague and potent distress. This distress is rarely deserved and is somehow in the nature of things; in later sentimental works it even overshadows virtue, which may in fact be more manifest in the sympathy of the observer than in the sufferer. The distressed are natural victims, whose misery is demanded by their predicament as defenseless women, aged men, helpless infants or melancholic youths. (2–3)

Animals, and particularly the dependent, devoted dog, render victimization in a higher, purer idiom. The dog serves as a voice of unmediated hence honest emotion; sensation, often pain, forms its very language. In addition, while Fred Kaplan argues that Victorian sentimentality was an antagonistic response to the "mechanical, or rational, or deterministic, or pragmatic forces" expressed as realism (6), dog narratives curiously interweave sentimentality with the experiential, flawed view of human nature associated with realism. Moreover, as changing conceptions of femininity threatened to unsettle the underpinnings of the most popular Victorian fictional form, domestic realism, with revolutionary resistance to the doctrine of separate spheres and its divisions of feeling and function, the canine protagonist provides an ungendered heroine. The dog narrator's gender, whatever it is, rarely signifies as important. Thus the dog provides a comforting substitution for the domestic heroine when she perfidiously questions the characteristics of which she is constructed: modesty, affection, submission, and loyalty. Finally, as Gillian Beer has convincingly argued, the rise of scientific interest in animals—particularly manifested in Charles Darwin's work—challenged belief in religious hierarchies which placed humans not as merely distinct from animals but as superior to them. While Christianity authorized humans to dominate and make use of animals, the conclusions drawn from science were more ambiguous. The possibility that animals—including humans—rose from the same origins was unsettling enough, but the period also produced studies that argued for an interior, psychological animal life—for animal subjectivity, in effect—such as Darwin's own *The Expression of the Emotions in Man and Animals* (1872) and George Romanes's *Animal Intelligence* (1883). One of the most fascinating consequences of the synthesis of these systems of thought and taste was dog narrative, a crystallization of powerful emotion, victimization, domestic displacement, and heretofore mute eloquence. Collectively, these characteristics not only produced the speaking dog but attached that figure to very human fears of passing time, changing values, loss, and even death. The old dog, not *the* dog, emerges as the canine voice of authority. Kathleen Kete finds similar attention to older dogs on the part of the French. As she briefly notes of Paris and as Harriet Ritvo similarly finds in Britain, stories frequently circulated of dogs visiting the graves of their dead masters, of dogs dying of heartbreak, even of dogs committing suicide when separated from beloved humans. Kete interprets these stories and the association of canine fidelity with death as a symptom of urbanity and modernity. The dog becomes at once the signifier of emptiness and a feeble attempt to ameliorate alienation (27). In the British context the dog narrator represents both less and more. The sheer diversity of examples argues that Victorians—anticipating mass consumers of books, television programs, and films today—came to depend heavily on anthropomorphic narratives. The emotions permitted the animal narrator allowed the losses with which late life is so often associated to be expressed. The aging animal invited compassion for the seldom examined but fearfully imagined years following an active, engaged, productive, and reproductive midlife, and also for that sadly anticipated moment after death in which the older person is forgotten.

I began work on this project because I wanted to understand how these narratives function as a polemic against brutal treatment of marginalized figures, in particular of older humans. I am slowly coming to believe that many of these narratives—and the imaginary animals who narrate—function unexpectedly as affirmations, albeit sentimental, of the value of "secondary" creatures, or at least of their stories. And these stories include narratives of those people whom the Victorians designated "the aged," especially the poor or infirm, living and dying in the nineteenth century.

It would be easy to attribute our pleasure in old dog narrators and narratives to a narcissism that can only really be sustained by animal silence. Into that silence we read our woes and our consolations. "In a world of hypocrites," wrote the romance novelist Ouida, only the dog can be counted upon for kindness: "For your dog you are never poor; for your dog you are never old; whether you are in a palace or a cottage he does not care; and fall you as low as you may, you are his providence and his idol still" ("Dogs" 318). When readers respond to the suffering, the unpretentious wisdom, and even, perhaps especially, to the sentimentality of Victorian old animals' fictive voices, are they in fact being compelled—however clumsily—by the possibilities of subjectivities we generally either deny or disregard (in this case the subjectivity of either animals or older people, or of both)?

In his essay "Killing with Kindness: Veterinary Euthanasia and the Social Construction of Personhood," veterinarian Clinton R. Sanders considers how he, his clients (pet owners), and other veterinarians perceive the euthanasia of pets—which happens most frequently when old animals become incapacitated or appear to be suffering. In principle, animals are put to death—as food, as public nuisances, as suffering creatures—with little compunction because they are not perceived as possessing selfhood or agency. At best they possess accumulated experience rather than "subjectivity." The pet-owner relationship, however, calls this dynamic into question for most people involved in the choice to euthanize a pet. As Sanders explains,

> The emotional intensity of the relationships that often develop between people and their nonhuman companion animals commonly prompts human caretakers to be ambivalent about, or reject entirely, the definition of their animals as mindless, objectified, nonpersons. . . . The emotional ambivalence of the key actors in the euthanasia encounter calls attention to the fact that the "personhood" of sentient others is a matter of social definition. Those "candidates" who are effectively excluded from the category of person may be routinely disposed of with minimal emotional cost. In contrast, the deaths of social others who are defined as minded, autonomous, and self-aware individuals with whom one may have authentic and emotionally rewarding mutual relationships typically precipitate intense grief. The elemental issue of the construction of social identity, I maintain, is most strikingly revealed through examining interactions with those who are defined as being on the border between person and nonperson. (197–98)

This consequent, constructed animal subjectivity is further complicated by the silence of the pet. In what Sanders calls a "triadic relationship," the "human client is commonly called upon to provide everyday 'interactional' information to supplement the 'technical' and 'perceptual' information elicited through the doctor's use of his or her training and equipment. Veterinarians and their clients cooperate to cast the animal as a 'virtual patient'" (200). In other words, pet owners—or caretakers—have long been "dog narrators"—accustomed to speak for, even to speak as, their old dogs.

Many of us who have pets love our pets in large part because they *cannot* talk. Why then do we take pleasure in the fantasy of a narrating animal? Why do we long to narrate what we feel convinced are their thoughts and feelings? I would argue that the link between animal age and animal narrator depends upon a paradox which may be a grim reminder that old age, like animal life, is a tableau that our culture prefers to see blind, silent, and bathed in sentiment. In the nineteenth century, "mad dogs"—which often meant any dog roaming the streets unmuzzled—were beaten, placed under observation, or shot. Little wonder that when mute or muted creatures speak, they speak in the voices of sentiment, suffering, abjection, or forbearance for the readers who must be persuaded to suspend cultural assumptions about any others, including humans deemed non-human, and suddenly submit to unimaginable, inarticulate selves.

Victorian novelists experimenting with animal narrators seem to be negotiating with a scenario the veterinarian Sanders poses. The fictional animal is a form of subjectivity that requires us not only to attend to the stories it tells but to reconceive our long-told narratives of helplessness, frailty, utility, and obligation as we struggle to comprehend marginal subjects. We are compelled to compose these alternative narratives in the face of our older human companions, of our aging selves, and of death itself—a subject seldom shirked by canine storytellers.

Notes

I wish to thank my research assistants Sean Scanlon and Marty Gould, who tracked many of my sources to their lairs. Thanks also to Lori Branch, Corey Creekmur, Eric Gidal, and Judith Pascoe for their astute suggestions.

1.   These are too innumerable to document, but one can get a flavor of the poetry across the century by sampling the anonymous "My Dog's Epitaph. By the Subaltern"; "Lines to the Memory of a Favourite Dog," which opens "Poor dog, and art thou dead!"; and Louise Imogen Guiney's "To a Dog's Memory."
2.   The photographs are reproduced in Gibson.
3.   Keith Thomas provides an excellent discussion of the ways in which sentimental attitudes (emotional rather than explicitly literary) arise in the seventeenth century in his *Man and the Natural World*.

Bibliography

"Animals in Novels." From *The Spectator*. Rpt. in *The Living Age*, 5th ser., 212 (Jan.–Mar. 1897): 411–14.

Beer, Gillian. *Darwin's Plots: Evolutionary Narrative in Darwin, George Eliot, and Nineteenth-Century Fiction*. Cambridge: Cambridge University Press, 2000.

Budiansky, Stephen. "The Truth About Dogs." *Atlantic Monthly* 184 (July 1999): 39–53.

Darwin, Charles. *The Expression of the Emotions in Man and Animals*. 1872. London: Pickering, 1989.

"A Dog on His Day." *Punch* 29 Dec. 1894: 302.

"Dogs of Literature." *Temple Bar* 61 (Jan.–Apr. 1881): 476–500.

"A Drunken Dog." From *The Saturday Review*. Rpt. in *Littell's Living Age*, 5th ser., 181 (Apr.–June 1889): 703–704.

Fitzgerald, William G. "Dandy Dogs." *Strand Magazine* 11 (Jan.–June 1896): 538–50.

Gibson, Robin. *The Face in the Corner: Animals in Portraits from the Collections of the National Portrait Gallery*. London: National Portrait Gallery Publications, 1998.

Guiney, Louise Imogen. "To a Dog's Memory." *Century Magazine* 38 (Oct. 1889): 947.

Harrison, Brian. *Peaceable Kingdom: Stability and Change in Modern Britain*. Oxford: Clarendon, 1982.

Kaplan, Fred. *Sacred Tears: Sentimentality in Victorian Literature*. Princeton: Princeton University Press, 1987.

Kete, Kathleen. *The Beast in the Boudoir: Petkeeping in Nineteenth-Century Paris*. Berkeley and Los Angeles: University of California Press, 1994.

Kipling, Rudyard. *Collected Dog Stories*. Garden City, N.J.: Doubleday, 1934.

Lansbury, Coral. *The Old Brown Dog: Women, Workers, and Vivisection in Edwardian England.* Madison: University of Wisconsin Press, 1985.

The Life of Carlo, the Famous Dog of Drury-Lane Theatre. 1806. London: Tabart, 1812.

"Lines to the Memory of a Favourite Dog." *Blackwood's Edinburgh Magazine* 22 (Oct. 1827): 439–40.

"My Dog's Epitaph. By the Subaltern." *Blackwood's Edinburgh Magazine* 19 (June 1826): 685.

"My Old Dog and I." *Blackwood's Edinburgh Magazine* 22 (Dec. 1827): 731–33.

Ouida. *A Dog of Flanders: A Christmas Story.* Boston: Page, 1891.

———. "Dogs and Their Affections." *North American Review* 153.418 (1891): 312–21.

"Page from a Dog's Diary." *Punch* 1 Nov. 1899: 207.

Ritvo, Harriet. *The Animal Estate: The English and Other Creatures in the Victorian Age.* Cambridge, Mass.: Harvard University Press, 1987.

Romanes, George John. *Animal Intelligence.* New York: Appleton, 1883. Rpt. in *Significant Contributions to the History of Psychology, 1750–1920.* Ed. Daniel N. Robinson. Ser. A. Orientations. Vol. 7. Washington: University Publications of America, 1977.

Sanders, Clinton R. "Killing with Kindness: Veterinary Euthanasia and the Social Construction of Personhood." *Sociological Forum* 10.2 (1995): 195–214.

Saunders, Marshall. *Beautiful Joe: An Autobiography.* Philadelphia: Griffith, 1893.

Secord, William. *Dog Painting, 1840–1940: A Social History of the Dog in Art.* Woodbridge, U.K.: Antique Collectors' Club, 1998.

Sewell, Anna. *Black Beauty: The Autobiography of a Horse.* 1877. New York: Grosset, 1945.

Shorter, Clement K. "Relics of Emily Brontë." *The Bookman: A Literary Journal* 6 (Sept. 1897): 15–19.

Stables, Gordon. *Sable and White: The Autobiography of a Show Dog.* London: Jarrold, 1894.

Thomas, Keith. *Man and the Natural World: A History of the Modern Sensibility.* New York: Pantheon, 1983.

Todd, Janet. *Sensibility: An Introduction.* New York: Methuen, 1986.

Tracy, Marguerite. "A Dog's View of Things." *The Bookman: A Literary Journal* 6 (Sept. 1897): 71–72.

Turner, E. S. "Animals and Humanitarianism." Rpt. in *Animals and Man in Historical Perspective.* Ed. Joseph Klaits and Barrie Klaits. New York: Harper, 1974. 144–68.

Wise, John Sargeant. *Diomed: The Life, Travels, and Observations of a Dog.* Boston: Lamson, 1897.

4 The Moral Ecology of Wildlife

Andrew C. Isenberg

I once challenged the students in my undergraduate seminar, "American Environmental History," to define capital-N "Nature." The participants in the seminar had read selections from the nineteenth-century wilderness advocate John Muir together with works by environmental historians critical of the concept of "wilderness."[1] The students largely accepted these critiques, which shared the premise that the term "wilderness" has no specific ecological meaning but rather denotes certain kinds of cultural significance (authenticity, freedom, virtue, and godliness, for instance) that Americans invest in certain landscapes. Once they had embraced environmental history's critique of the notion of wilderness, however, my students struggled to reconstruct a workable definition of Nature. Finally, after considerable discussion, and not without a certain exasperation with the whole exercise, one of the students blurted out, "'Nature' is where Bambi lives." The assembled students, perhaps weary of the direction of discussion or suspicious, as undergraduates will be, that the professor had asked one of those questions to which there was no real answer, readily assented to this formulation.

Without realizing it, they had recapitulated twentieth-century Americans' struggle to redefine Nature beyond the scenic but static nineteenth-century concept of "wilderness." In the nineteenth century, preservationists drew borders around scenic landscapes such as Yosemite and Yellowstone and invited urbanites to experience the restorative effects of Nature. Wildlife within those borders was an afterthought; Congress passed effective legislation against poaching in national parks only in 1896, and did not begin to restock parks with species such as bison until the first decade of the twentieth century. Even then, the bison were corralled, semi-domesticated spectacles. Change came in the twentieth century as Americans gradually came to the consensus that, to be authentic, Nature must teem with life—*wild*life, to be precise. This consensus represented a trickling down of the work of scientists such as Charles Elton, A. G. Tansley, and Eugene Odum, who, beginning in the 1930s, emphasized that the natural environment's energy and material was in constant flux. The popularization of ecology in the 1960s brought such notions into the mainstream.

But the formulation "'Nature' is where Bambi lives" captures not only the scientific but the sentimental. Sentimentalization of farm animals and pets was common in the nineteenth century, as an industrializing society romanticized its rural past, but an emotional regard for wild animals is a distinctly twentieth-century (or, at least, a late-nineteenth-century) phenomenon. It was first exemplified in the beginning of the century in the writings of Ernest Thompson Seton and Jack London; it was reformulated and informed by science and ethics by Aldo Leopold at mid century; and it was encapsulated in legislation in the Endangered Species Act of 1973 and in concerted efforts in the 1990s to reintroduce animals from the endangered list to their former habitats.

The changing apprehension of wildlife in American culture is perhaps best exemplified in the long journey that Bambi's primary predator, wolves, have traveled through the American imagination. Feared and reviled as loathsome and cowardly killers at the outset of the century, they have come, by the century's end, to symbolize the possibility for holism and integrity not only in the American environment, but in American culture. The reintroduction of wolves to certain environments in the 1990s represented not just a scientific effort to reconstruct functioning ecosystems by restoring an important predator to its place in the food chain. It also represented an assertion of a moral order, a belief in the inherent integrity of an ecosystem managed not by people but by wildlife. In a departure from earlier ideas about the management of Nature, twentieth-century wolf advocates argued that restoring a vital, functioning Nature would come at a price: wolves would prey on domesticated livestock and perhaps even on household pets. An authentic Nature, they argued, required sacrifice. But that sacrifice would restore vigor and authenticity to human society. Wilderness advocates in the nineteenth century had advanced a similar idea. They had argued that urban society was corrupt and debilitating, while unspoiled wilderness was pure, and that the experience of wilderness was an antidote to pampered cosmopolitan life. But in nineteenth-century wilderness ideology, the distance between what was human and what was natural was unbridgeable. In the twentieth century, Americans began to imagine that they could close that gap.

I

In the second half of the nineteenth century, most Americans despised, or at best ignored, wolves and other wildlife. Whatever concerns Americans gave to animals they expended on pets and farm animals. In the years after the Civil War, urban, middle-class reformers turned their energies from abolition and temperance to animal protection. Between 1866 and 1874, the American Society for the Prevention of Cruelty to Animals (ASPCA) established chapters in thirty of the largest cities in the Northeast and Upper Midwest. Critical and fearful of urbanization and industrialization, the ASPCA romanticized rural America and what they presumed to be the close, caring contact between farmers and their animals. If middle-class children would keep pets, and carriage drivers would treat their draft horses with the sort of kindness that farmers presumably lavished on their sheep and dairy cows, the ASPCA argued, then rural values could find a place in the competitive amorality of the industrial city (Turner 31–57).

The ASPCA was unable, however, to prevent the wasteful hunting of wildlife in the second half of the nineteenth century. For the general public, the ASPCA's rhetoric of kindness applied only to tame animals. The organization failed, for instance, in its effort to halt the destruction of the bison in the 1870s, a slaughter that exemplified the nineteenth-century commercial destruction of wildlife. The salvation of the bison from extinction came only between 1905 and 1914, when an organization of nostalgic naturalists and hunters, the American Bison Society, engineered the creation of several bison preserves. Much as the ASPCA had rued the loss of rural innocence in an urban age, the Bison Society feared the extinction of an icon of the bygone frontier. The members of the Bison Society had apprehensions of endangerment, but they feared less the extinction of the species itself than the extinction of the masculine frontier culture that bison hunting exemplified (Isenberg 143–56, 164–85).

One of the members of the American Bison Society was the mercurial Ernest Thompson Seton, best known as one of the founders of the Boy Scouts of America. Born Ernest Evan Thompson in England in 1860, he emigrated with his family to Canada in 1866. Like other late-nineteenth-century romantics, he asserted his distant noble lineage; in his case, he took the surname of the Scottish Jacobite House of Seton in 1883. After failing as a painter and naturalist in the 1880s and early 1890s, Seton became wealthy by writing and illustrating a series of stories about wild animals. A sample of the titles gives a sense of Seton's work: *Wild Animals I Have Known* (1898), *The Biography of a Grizzly* (1900), *Animal Heroes* (1901), *Lives of the Hunted* (1901), and *Wild Animals at Home* (1917). While the stories made him rich, contemporary naturalists scorned his sentimental, anthropomorphized animal stories as deliberate misrepresentations of wildlife.[2] John Burroughs, the acclaimed nature writer, attacked Seton as a "nature faker" in the *Atlantic Monthly* in 1903:

> In Mr. Thompson Seton's *Wild Animals I Have Known* . . . I am bound to say that the line between fact and fiction is repeatedly crossed. . . . Mr. Thompson Seton says in capital letters that his stories are true, and it is this emphatic assertion that makes the judicious grieve. True as romance, true in their artistic effects, true in their power to entertain the young reader, they certainly are; but true as natural history they as certainly are not. (qtd. in Keller 153–54)

Burroughs is, of course, quite correct. Seton's stories (of Silverspot the Crow, Raggylug the Cottontail, and Redruff the Partridge, for instance) are inventions that are merely "true as romance." Seton used the literary tools available to him at the time, borrowing from nineteenth-century genres—late Victorian moral lessons and popular, sensationalistic adventure writing—to animate nature. This was not strictly the dynamic nature of Charles Darwin, the "struggle for life," largely unthinking and entirely without moral consequence, in a bloody arena of competition. Seton's nature could be quite bloody, but by investing wildlife with human attributes—emotion, reason, and most importantly conscience—Seton made nature resonate with the concerns and choices of humanity. In other words, Seton's animals, if anthropomorphized and therefore unreal, were nonetheless "authentic" in the sense that they were apt projections of modern anxieties, particularly the desire to transcend an overcivilized, artificial existence and experience real, even primal, sensations.[3]

One of Seton's most poignant stories, contained in his 1898 collection *Wild Animals I Have Known,* is of Lobo, a cunning, powerful wolf who inhabited the Currumpaw rangelands of northern New Mexico. Lobo inspired terror in cattle and despair in ranchers. Guns and traps were hopelessly ineffective against him. He and his pack ate only what they had killed themselves, in order to avoid poisons and traps. In Seton's tale, it is Lobo's humanity that finally dooms him. Seton, unable to trap Lobo, lures Lobo's mate, Blanca, into a clever snare, then drags Blanca's body through a narrow pass that he has mined with traps. Lobo, reckless in his search for his beloved mate, is snared (*Wild Animals* 17–54).

In one sense, Seton reflected the perspective of the nineteenth century, which sought the destruction of most wildlife, particularly predators, in order to domesticate the environment: he unapologetically killed Lobo and Blanca in order to protect Currumpaw livestock. Yet his characterization of Lobo and his pack heralded a new representation of wildlife. Seton's wild animals inhabit a moral universe of honor, love, and choice. Seton regards Lobo not as a varmint but as a worthy adversary. Lobo must die because he preys on livestock, but Seton's admiration for him is obvious. Lobo is a sort of noble savage. Indeed, in Seton's stories and those of Jack London, wildlife in general and wolves in particular assume the literary persona of the Indian, the noble savage.

By Seton's time, nineteenth-century dime novelists had thoroughly developed the noble savage persona. The Indians in these formulaic stories possess, however incongruously, superhuman skills in hunting, pathfinding, and war; a childish naiveté and temperament; and a primitive sagacity (see Brown 1–40; and Smith). Seton's wolves (like those in Jack London's writings) likewise possess this unlikely combination of traits. Like the dime novelists, Seton and London took the most "savage" contemporary subjects for their work. The dime novelists wrote of Indians during the mid to late nineteenth century, while numerous native groups remained autonomous in many parts of the American West. In the wake of the Indians' subjugation, the wildlife writers of the early twentieth century neatly transferred the dime-novel Indian's dubious qualities to certain animals.

The literary persona of the dime-novel Indian/animal-story wolf is a conflicted one. It is formed by equal parts of admiration and revulsion. The historian Philip Deloria has argued that this contradiction is at the heart of the concept of the "noble savage." Noble savagery, Deloria wrote in 1997,

> both juxtaposes and conflates an urge to idealize and desire Indians and a need to despise and dispossess them. A flexible ideology, noble savagery has a long history, one going back to Michel de Montaigne, Jean-Jacques Rousseau, and other Enlightenment philosophers. If one emphasizes the noble aspect, as Rousseau did, pure and natural Indians serve to critique Western society. Putting more weight on savagery justifies (and perhaps requires) a campaign to eliminate barbarism. Two interlocked traditions: one of self-criticism, the other of conquest. They balance

perfectly, forming one of the foundations underpinning the equally intertwined history of European colonialism and European Enlightenment. (4)

Likewise, for Seton, Lobo may be a pillaging freebooter, but he is blessedly free of cosmopolitan neuroses. By hunting and killing the noble savage, Seton, by implication, is ennobled (Slotkin 1–62). Historians of the American West are all too familiar with this narrative trope, whose primary advocates in the late nineteenth century were the historians Frederick Jackson Turner and Theodore Roosevelt. Richard Slotkin termed it "regeneration through regression": in the process of conquering the wilderness, the conqueror is first de-civilized and then reborn as a better person.

The ennobling effect of wildlife pervades the work of another early-twentieth-century writer, the novelist Jack London, who was born in California in 1876, raised in poverty, and schooled in vagrancy, piracy, and (briefly) the University of California at Berkeley. Like Seton, London anthropomorphized animals. His most widely read novella, *The Call of the Wild,* is a thinly disguised autobiography, with the author, who joined the gold rush to Alaska in 1897, as Buck, a St. Bernard–shepherd mix abducted in California and sold as a sled dog in Alaska. Adhering to the terms of the "regeneration through regression" genre, Buck is toughened up by the rigors of hauling sleds on the Alaska frontier. But Buck's devolution from civilization does not stop there. When Indians kill his master, Buck goes feral and joins a pack of wolves. London, unlike Seton, therefore violates the cardinal rule of the "regeneration through regression" genre, and in doing so transcends its clichés: rather than returning to civilization reinvigorated by the encounter with wilderness, Buck regresses all the way to permanent savagery. What makes this complete regression remarkable is that London, unlike most turn-of-the-century thinkers about wildlife, presents this regression as a triumph. One way to read *Call of the Wild,* therefore, is as a parody of the late Victorian moral improvement tale. Like the protagonists of Horatio Alger's novellas, who progress from rags to respectability (see Trachtenberg v–xx), Buck learns a series of lessons under the discipline of the wilderness. He finally achieves a kind of moral redemption by shaking off the last vestiges of civilization and joining a wolf pack.

II

In the early twentieth century, conservationists were reluctant to embrace the idea that wildlife possessed a redeeming power. The mission of Progressive-era conservationists was, rather, the utilitarian management of natural resources. In creating the National Forests in 1905,[4] conservationists were responsible for withdrawing millions of acres of forest- and rangelands from the path of unregulated economic development. The conservationists' goal was to rationalize land use, to manage natural resources to maximize sustainable production.

If this mission meant preventing cut-and-run lumbering in order to save timber companies from themselves, however, it also meant a campaign to exterminate certain predators in order to make rangelands safe for livestock. The Bureau of Biological Survey, created, like the Forest Service, in 1905, had as one of its mandates the destruction of varmints. By 1907, the Bureau was responsible for the deaths of 1,800 wolves and 23,000 coyotes in the National Forests. By 1931, three-quarters of the Bureau's budget went to the professional hunters in the predator-control program. Between 1915 and 1942, hunters killed over 24,000 wolves (Worster, *Nature's Economy* 262–64; Lopez 187). Under the auspices of the program, hunters killed the last wolf in Yellowstone Park in 1926, and the last in southwestern Montana in 1941 (Altenhofen). Stanley P. Young, a senior biologist at the Department of the Interior, summarized the reasons for the campaign against the wolf in 1944: "The wolf was not only a menace to human life, but was everywhere so destructive to domestic stock that constant warfare had to be waged against it" (Young and Goldman 1).

The primary goal of predator extermination was the protection of domesticated livestock; an ancillary benefit, it was thought, was to boost the population of game animals. In the early years of the century in the Kaibab National Forest in Arizona, federal predator control sought to increase the population of deer. In one of his most poignant essays, "Thinking like a Mountain," Aldo Leopold, a graduate of the premier training ground for conservationists, the Yale School of Forestry, remembered his days as a forest ranger hunting wolves in the Southwest: "I was young then, and full of trigger-itch. I thought that because fewer wolves meant more deer, that no wolves would mean hunters' paradise" (130). In other words, at this stage in his career, Leopold's definition of nature was not unlike that of my undergraduates: it was where Bambi lived.

Because of the labors of hunters to remove predators other than human, the deer population of the Kaibab National Forest mushroomed from 4,000 in 1906 to 100,000 in 1924. In the next two years, however, unable to find sufficient forage, 60,000 deer starved. By 1939, the Kaibab deer population had declined to 10,000 (Worster, *Nature's Economy* 270–71).[5]

To naturalists of the 1930s, the Kaibab deer irruption was evidence that even trained conservationists could not manipulate the environment to utilitarian ends—that to remove one part of nature because it interfered with people's productive use of the land damaged the intricate web of interconnections that make up what came to be called, in a 1935 essay by the British ecologist A. G. Tansley, the "ecosystem" (Worster, *Nature's Economy* 301–302).

Leopold was likewise dismayed by the Kaibab disaster. Like St. Paul on the road to Damascus, he dated his conversion from conservationism to environmentalism—a distinction I'll clarify in a moment—from his experience as a wolf-killer. He wrote of his participation in the death of a wolf:

We reached the old wolf in time to watch a fierce green fire dying in her eyes. . . . Since then I have lived to see state after state extirpate its wolves. . . . I have seen every edible bush and seedling browsed [by deer], first to anaemic desuetude, and then to death. I have seen every edible tree defoliated to the height of a saddle-horn. . . . [W]hile a buck pulled down by wolves can be replaced in two or three years, a range pulled down by too many deer may fail of replacement in as many decades. So also with cows. The cowman who cleans his range of wolves does not realize that he is taking over the wolf's job of trimming the herd to fit the range. . . . Perhaps this is behind Thoreau's dictum: In wildness is the salvation of the world. (131–33)

As attractive as it may be to imagine that Leopold experienced a moment of moral clarity when he looked into the eyes of a dying wolf, his rejection of predator-control programs likely occurred after he left the Southwest for Wisconsin in 1924, where he eventually took up a position in the Department of Game Management at the University of Wisconsin at Madison. The situation in Wisconsin in the 1930s and 1940s mirrored that in the Kaibab: thousands of deer were starving while the state government persisted in its wolf eradication program. Leopold was closely involved in the effort to lift the state's wolf bounty (Theil 87–112).[6]

Leopold was not the only scholar in the first third of the twentieth century who looked to North American natural history for an answer to the problem confronting game managers. As early as the 1910s, the leading authority on the native hunting groups of Canada, the anthropologist Frank Speck, argued that Indians' sustainable hunting had functioned to conserve game supplies ("Family Hunting Band"; and "Mistassini"). Leopold, like Speck, argued that wild fluctuations of game populations, whether "irruptions" of deer populations or the near-extinction of the bison, were the consequences of Euro-Americans' disruption of the harmony that had existed in Precolumbian North America among predators (both human and non-human) and their prey.

In his parable[7] of the dying wolf, Leopold asserts that animals—even those, like wolves, widely regarded as varmints—were living embodiments of what one might call the "moral ecology of wildlife."[8] This term is inspired by the historian E. P. Thompson's work on "moral economy." Thompson's 1971 essay "The Moral Economy of the English Crowd" located the source of eighteenth-century food riots not in mere hunger but in the rioters' assertion of traditional rights. Similarly, in late-seventeenth- and early-eighteenth-century England, some of the strongest assertions of the customary rights associated with traditions of moral economy involved deer hunting. Gamekeepers had long looked the other way when local individuals poached deer on royal forests. In doing so, they deferred to the poachers' claims to customary rights to hunt for subsistence. After 1723, when Parliament sought to put an end to the moral economy of hunting by making poaching a capital crime, the poachers blacked their faces, hunted at night, violently confronted gamekeepers, and began to construe poaching as an overtly political act of defiance of authority ("Moral Economy"; and *Whigs*).[9]

We can discern here a consistent pattern of thought. Thompson argued that the predation of blackfaced hunters on royal game was a forceful assertion of a traditional order in the face of aristocrats' efforts to reserve resources for themselves. According to Leopold, the predation of wolves on domesticated livestock was a forceful assertion of nature's moral economy on Americans' efforts to manipulate the environment for their own benefit. At first glance, it may seem as if a great interpretive chasm separates E. P. Thompson and Aldo Leopold. Yet a cultural thread, stretched and frayed but definite, connects the tradition of moral economy with what I have called Leopold's moral ecology.

According to Deloria and other historians, the English tradition of "misrule" was "Indianized" in late-eighteenth-century America by colonists searching for an indigenous claim to the tradition of moral economy. Rather than blackface, Americans masqueraded as Indians by daubing their faces with ersatz warpaint. For Anglo-American colonists, "playing Indian" meant more than symbolically distancing themselves from England. Eighteenth- and early-nineteenth-century Americans also attributed qualities of liberty, independence, and natural rights to Indians. Thus, New Hampshire loggers resisting the Mast Tree law that reserved valuable white pines for the Royal Navy, the Sons of Liberty in Boston protesting import duties on tea, the protesters in the Whiskey Rebellion in western Pennsylvania in the 1790s resisting taxes on alcohol, and the backcountry tenants in Maine protesting rents levied by their landlords: all disguised themselves as Indians to assert what they regarded as their customary rights according to borrowed (English) and invented (Indian) traditions of moral economy (Deloria 10–37; see also Slaughter; and Taylor).

Seton and London had transferred the literary persona of dime-novel Indians to wildlife; likewise, Leopold tapped into the American tradition of Indianized moral economy and transformed it into the moral ecology of wildlife. Moral ecology went beyond the mere sentiment of Seton. It assumed, like Seton and those who followed him, such as Felix Salten (the author of *Bambi: A Forest Life*) and Walt Disney (the producer of the 1941 animated film version of *Bambi*), that wild animals inhabit a moral universe and that people would do well to emulate the innate morality—the natural law—of the wild (see Cartmill). It goes beyond them in asserting that the order of nature constitutes a higher, morally and scientifically integrated order. In this regard, it resembles Speck's work on Indian hunters, which argued that the "savage" was a better conservationist than the civilized Euro-American.

Seton's wolf, Lobo, was a bandit. For Seton, consumed by the turn-of-the-century fear that civilized comforts were dulling the senses, Lobo was a colorful pirate, a refreshing, even admirable antidote to the ills of urban industrial society. But Lobo was an outlaw nonetheless, and therefore doomed to be destroyed by what Seton regarded as the inevitable progression of the laws of property and profit. By contrast, Leopold's wolf was, like Leopold himself, a kind of forest ranger, enforcing the moral ecology of wildlife. Seton had to regress from civilization to encounter wolves. Leopold saw in wolf ecology a higher law. Yet despite this crucial difference it is hard to imagine Leopold's wolf, with the "fierce green fire" in her eyes, without the precedent of Seton's Lobo. Seton made wolf ecology a world of conscience and reason; Leopold invested that world with moral purpose.

One such purpose, of course, was an implicit critique of capitalism. Rather than manipulate Nature for human ends (the essence of conservationism), Leopold argued that people should strive to adjust to the dictates of Nature (the essence of environmentalism). Seton and London depicted wolves as the antidote to cosmopolitan neurosis. In Leopold's "Thinking like a Mountain," wolves are the antidote to the capitalist degradation of nature. Leopold's critique of capitalism is grounded in the ecological and economic disasters of the 1920s and 1930s: Kaibab, the southern plains "dust bowl," the deer irruption in Wisconsin, and the Great Depression.[10] In the 1940s, Leopold elaborated these insights into an environmentalist credo, the "land ethic": "A thing is right when it tends to preserve the beauty, stability, and integrity of the biotic community. It is wrong when it tends otherwise" (224–25). Conservationists, in short, sought to save Nature. Environmentalists such as Leopold looked to Nature for humanity's salvation.

III

Despite Leopold's poignant expressions of regret for the killing of wolves, and his assertion of an ethical obligation to preserve the "biotic community," predator eradication programs continued well into the second half of the twentieth century. Wisconsin paid bounties for wolves until 1957, Michigan until 1960, and Minnesota until 1965. Alaska, with a wolf population so large that the species was never listed as endangered in the state, persisted in its wolf eradication program until 1995 (see "Alaska Governor"; and Line). By the 1970s, while thousands of wolves survived in Canada and Alaska, the wolf population in the lower forty-eight states had been reduced to a few hundred in the remote, largely roadless forests of northeastern Minnesota.

One of the enigmas of twentieth-century American environmental history is explaining how, in this context of culturally sanctioned wildlife decline, a sudden flurry of environmental legislation protecting wilderness was enacted: the Wilderness Act of 1964; the National Wild and Scenic Rivers Act of 1967; the Marine Mammal Protection Act of 1972; and three successive Endangered Species Acts in 1966, 1969, and 1973. Historians' explanations for this spasm of activity focus on the increasing view of wilderness as a valuable consumer amenity during the period of postwar prosperity, and the pressure of relatively wealthy and well-organized wilderness activists in a legislative and regulatory system driven by interest group politics (see Nash; Hays, *Beauty*; Gottlieb; Sale; Shabecoff; and Rothman). Certainly, sentimental popular interest in wildlife was high: according to the historian Thomas Dunlap, the national broadcast of a wildlife documentary in November 1969, "The Wolf Men," which graphically portrayed the trapping and killing of wolves, prompted a wave of public indignation (see Dunlap, *Saving* 148–49).

Public interest, however, did not dictate the form or philosophy of wildlife preservation; these were rooted in the notion, sometimes unarticulated or even unrecognized, of moral ecology. The language of moral ecology is embedded in the wildlife legislation of the 1960s and 1970s, put there by a generation of wild-life biologists and wilderness activists trained in Leopold's treatises on game management and injunctions to preserve nature. The Marine Mammals Protec-tion Act sought to prevent the populations of seals, whales, and other species from declining "beyond the point at which they cease to be a significant func-tioning element of the ecosystem of which they are a part" (*Statutes* 86: 1027). Likewise, the Endangered Species Act of 1973 intended to do more than stave off the extinction of discrete species. Rather, like the act protecting marine mammals passed a year earlier, the legislation was conceived at a systemic level: "The purposes of this Act are to provide a means whereby the ecosystems upon which endangered species depend may be conserved" (*Statutes* 87: 885). The Endangered Species Act is thus designed to do more than save remnant popu-lations (which is all that turn-of-the-century preservation aimed to do). It is designed to reestablish wildlife in its role in regulating the operations of Nature. Unlike the late nineteenth- and early-twentieth-century wilderness movement, which bounded Nature inside the national parks, much of the wildlife protec-tion of the 1960s and 1970s knew few bounds. The ecosystems that endangered species were charged with maintaining were not segregated in remote places but included human beings and their domesticated animals. Ironically, the Endan-gered Species Act is a human ordering of the environment that at some level presumes that a human-ordered environment, as distinguished from one po-liced by wolves or other non-human predators, is an environment in distress.

Because wolves are top predators, biologists regarded them as important managers of ecosystems. In 1973, the gray wolf was the first species to be listed under the Endangered Species Act. The act called not only for the protection of animals from harm, but for "habitat acquisition and maintenance, propagation, live trapping, and transplantation" (*Statutes* 87: 885). The initial effort to ex-pand the range of wolves beyond Minnesota was an utter failure, however. Within months of their introduction to Michigan in 1974, four Minnesota wolves were killed by hunters or trappers. While Reagan-era conservatism sty-mied further efforts at reintroduction in the 1980s, the wolf population in the Upper Midwest and in parts of the Mountain West expanded without human intervention. Minnesota's wolf population doubled between 1979 and 1995, ris-ing to 2,200 and expanding its range to cover half the state. By the mid 1990s, Minnesota wolves had diffused into Wisconsin and northern Michigan; by the end of the century there were about one hundred wolves in each state. A like number of Canadian wolves migrated southward into western Montana (see Wiese et al.; Stevens, "Wolf's Howl"; Line; and McNamee).

Reintroduction did not begin in the Mountain West until January 1995, when twelve wolves captured in Alberta were released into Yellowstone and central Idaho. As in Michigan twenty-one years earlier, illegal killings began immedi-ately: one of the Idaho wolves was shot and killed two weeks after reintroduc-tion; the alpha male of one of the Yellowstone packs was shot and killed seven months after reintroduction. Nonetheless, the introduced packs thrived. By the summer of 1997, there were an estimated hundred wolves in Yellowstone (see

Johnson; "Wolves Arrive"; "Gray Wolves"; "Four Gray Wolves"; "Wolves in Yellowstone"; and Stevens, "As the Wolf").

Biologists asserted the benefits of wolf predation in the Yellowstone ecosystem: wolves would have a "pruning effect" on elk herds, eliminating the sick and old; predation would ease the pressure on forage overgrazed by the elk; wolf kills would create carrion for scavengers such as eagles and foxes. Many of these predictions were accurate. A study conducted two years after the first release found that the wolves had killed half of the coyotes in Yellowstone, causing a steep rise in rodent populations, which in turn benefited hawks and eagles. In short, the reintroduction increased biodiversity and improved the integrity of the Yellowstone ecosystem (Stevens, "Wolf's Howl" and "Triumph and Loss"; and Robbins).

Many biologists assume that the reintroduction of wolves has restabilized, or will soon restabilize, what they call the Yellowstone ecosystem. This assumption points to the dangers inherent in moral ecology's tendency to assume that wildlife's role in Nature is to preserve a rough harmony. Stability is a profoundly ahistorical notion. It presumes that, absent human influence, nature tends toward timelessness. An increasing number of biologists, however, believe that nature is inherently dynamic, and that stability is only an illusion. The naturally occurring fires in Yellowstone in 1987, which vastly altered the ecology of the park, demonstrated this kind of environmental dynamism. A recent study of the relationship between wolves and moose on Isle Royale in Lake Superior indicates that the relationship between the population of wolves and the population of their prey that undergirded Leopold's notions of ecology is probably far more complex than the tidy linear correspondence Leopold imagined. On Isle Royale, unpredictable ecological factors, particularly blizzards and the outbreak of zoonotic disease, seem more important in determining population (Peterson 145–63). Likewise, in the northeastern United States, wolves have confounded the categories of endangered species. Wolves migrating south from Canada have not displaced coyotes as in Yellowstone, but interbred with them. The problem of categorizing the offspring of a wolf (an endangered species) and a coyote (a non-endangered species) has perplexed policy-makers (Stevens, "Wolves May"). In short, Nature itself is an agent of historical change, an agent that the Endangered Species Act and the notion of moral ecology, grounded as they are in a concept of Nature that attributes all historical change to human agency, do not recognize.

The influence of Seton as well as Leopold was evident in biologists' assessment of the wolf reintroduction program. Scientists and advocates who favored reintroduction were well aware that they needed to cultivate popular support for the program. At an international conference on wolf ecology in Wilmington, North Carolina, in 1975, at a time when reintroduction was first being considered, a panel of scientists urged, "In public relations work, the wolf himself is our ally if we are minded to use him, his 'romantic' figure appealing to most people." In the discussion that followed, a scientist opined that "if Coca Cola can be sold to the public through television advertising, so can the wolf." Just as Seton trumpeted his role as a wolf hunter, wolf biologists understood that they could popularize the romantic aspects of their profession. "Our technical work is sometimes spectacular," wrote two scientists, "and our dedication to the wolf and ability to capture and handle the animals all serve to capture public interest and create a solid background for our factual information program" (Boitani and Zimen 475).

It would be wrong to conclude that most wolf biologists and wildlife advocates were cynically attempting to manipulate public perception by emphasizing romantic depictions of wolves. Seton's brand of anthropomorphism so colored their perspectives that they hardly needed to alter their understanding of their work to make it attractive to the public. For instance, without a trace of irony, the wolf advocate Rick McIntyre dedicated his book, *War against the Wolf,* to "Lobo and Blanca, Together in Life, Together in Death" (3). One National Park Service biologist described wolves released from their pens in Yellowstone as "cavorting, playing, and checking things out," a behavior that "suggests recent liberation" ("Wolves Leave"). Like Seton, wolf biologists invested the animals they studied with personalities; they followed matings and pack formations like a soap opera. Ten and Nine, the alpha male and female, respectively, of the Rose Creek pack in Yellowstone, first mated in their holding pens; like Lobo wooing Blanca, Ten had to coax the pregnant Nine out of the pen when the wolves were released; when Ten was shot and killed, Eight, a yearling male from another pack, brought food to Nine and her pups, and eventually became alpha male of the pack; meanwhile, Eight's old pack was displaced by a newer, more aggressive pack, led by its alpha male, Thirty-Eight. One biologist described Thirty-Eight's pack as "a bunch of hoodlums" ("Wolves Leave").

The sentimental pervades a 1998 novel about wolf reintroduction, *The Loop,* by the popular author Nicholas Evans. *The Loop,* which one might read as a rejoinder to Seton's romanticization of wolf hunting, tells the story of a conflict between ranchers and wolf biologists over a reintroduced pack in western Montana. The ranchers (all of whom are men) are unregenerate evil-doers, alienated from Nature, their wives and children, and their own emotions. The leading rancher, Buck Calder, a macho hunter and former rodeo star, is a bully and compulsive adulterer who psychologically terrorizes his wife and teenage son, reducing her to a recluse and him to a stutterer. Calder is surrounded by a toadying, all-male retinue of ranch hands and professional hunters whose emotions and outlooks are as straitened as his own. His compulsion to dominate extends to Nature, which is why he cannot abide the thought of wolves on the public lands where he owns grazing permits, particularly when the wolves begin to prey on his calves. The wolf biologists are the heroes of the novel. They, too, are flawed characters, but their flaws are the sorts of quaint neurotic tics that result from urban living. They are emotionally redeemed and made whole by their study of wolves. Helen, one of the wolf biologists, leaves a failed relationship behind in the East when she comes to Montana to study wolves. There, inspired by the wildlife, she recovers her spiritual and psychological health. She remarks, "[A]nimals were infinitely more reliable than people. . . . With their devotion and loyalty to each other, the way they care for their young, they seemed superior to people in almost every respect" (53). Likewise, Calder's son, Luke, an aspiring wolf biologist, contrasts his father's psychological abuse with the caring behavior of a male wolf, who, returning from hunting, lovingly retches up meat for his pups. For Evans, wolf reintroduction ministers not only to the health of the environment but more importantly to the human psyche. In his story of Lobo, Seton projected human emotions onto wolves. A century later, Evans's wolves lead full emotional lives; his human characters must tap into the wolves' emotions to fill their own emptiness.

Preservation of endangered species does not represent transcendent values, but reflects its historical context and legacy. The influence of Seton and Leopold pervades the Endangered Species Act, which justifies the preservation of fish, wildlife, and plants because of their "esthetic, ecological, educational, historical, recreational, and scientific value to the Nation and its people" (*Statutes* 87: 884). With the possible exception of "ecological," the values that justify the preservation of endangered species are human values.

Wildlife ennobles us; it may even be our salvation. But its values are those we ascribe to it, and the benefits of preserving wildlife accrue to us, too. Like the preservation of the bison, an example frequently invoked in public commentary on the preservation of endangered species ("Vanishing"; "Protecting"), the preservation of endangered species served particular cultural purposes (just as the effort to exterminate certain wildlife earlier in the century had served particular cultural purposes). There is no way around this conundrum. Our representations of wildlife are inescapably expressions of human values. Those values are historically contingent and inextricably entangled in a changing culture. We have, during the century just ended, decided that wildlife must be a part of Nature. But exactly what wildlife means to us, and why it must be a part of Nature, are questions that have not been answered. And, as my students suspected years ago, there may be no final answer.

Notes

1. The students read John Muir, "Wind-Storm in the Forest" and "Sequoia," in *Wilderness World;* Schmitt; Egan, "Trees" and "With Fate"; Matthiessen; and Baum. A more recent critique of wilderness is Cronon, "Trouble."
2. On Seton, see Anderson. For an excellent discussion of Seton's work, particularly the debate between Seton and John Burroughs, see Lutts.
3. For the problem of unreality and the search for authentic experience, see Lears, who argues that the search for authenticity was expressed particularly in historical romances of the 1890s (103–107).
4. The national forest system originated in the Forest Reserve Act of 1891 and the Forest Management Act of 1897. In 1905, under the Forest Transfer Act,

the forest reserves were transferred from the General Land Office to the Department of Agriculture and placed under the management of a professional Forest Service, under the direction of the chief forester, Gifford Pinchot (see Hays, *Conservation* 39–44).

5.   This is not to imply that, without human interference, deer populations would remain stable. Ungulate populations in particular are prone to irruptions and crashes (see Leopold, Sowls, and Spencer; Klein).

6.   For the gradual shift in Leopold's views, see Dunlap, "Aldo Leopold."

7.   William Cronon argued that "Historical wisdom usually comes in the form of parables, not policy recommendations or certainties" ("Uses" 16).

8.   The historian Karl Jacoby and I, working independently of each other, have both drawn on the work of E. P. Thompson to construct definitions of "moral ecology." For Jacoby's use of this term, see his cogent analysis of the impact of turn-of-the-century conservation law on rural America (3).

9.   My reading of the tradition of moral economy is inspired by Deloria (10–37). See also Scott, who explains peasant rebellions in Southeast Asia according to the peasantry's traditional, premarket social values that sought not to maximize profit but to minimize the risk of hunger.

10.   Historians writing in the 1930s, such as Walter Prescott Webb in the United States and Fernand Braudel in France, emphasized the limitations that geography and climate impose on human societies. For the 1930s, see Worster's *Dust Bowl.*

Bibliography

"After Being Released in Idaho, Wolf Is Shot to Death on Ranch." *New York Times* 31 Jan. 1995: A12.

"Alaska Governor Halts Wolf-Killing Program." *New York Times* 5 Feb. 1995: A29.

Altenhofen, Kelly J. "Shepherd of the Buffalo: The Wolf and Its Influence on Bison Movement and Mortality." Paper presented at "Bison: The Past, Present, and Future of the Great Plains," a Center for Great Plains Studies Interdisciplinary Symposium. University of Nebraska, Lincoln. 7 Apr. 2000.

Anderson, H. Allen. *The Chief: Ernest Thompson Seton and the Changing West.* College Station: Texas A&M University Press, 1986.

Baum, Dan. "Great Yellowstone Burn Now Seen as Gift from Nature." *Chicago Tribune* 17 July 1990: C1+.

Boitani, Luigi, and Erik Zimen. "The Role of Public Opinion in Wolf Management." *The Behavior and Ecology of Wolves.* Ed. Erich Klinghammer. New York: Garland, 1979. 471–77.

Braudel, Fernand. *The Mediterranean and the Mediterranean World in the Age of Philip II.* 1949. New York: Harper, 1972.

Brown, Bill, ed. *Reading the West: An Anthology of Dime Westerns.* Boston: Bedford, 1997.

Cartmill, Matt. *A View to a Death in the Morning: Hunting and Nature through History.* Cambridge, Mass.: Harvard University Press, 1993.

Cronon, William. "The Trouble with Wilderness; or, Getting Back to the Wrong Nature." *Environmental History* 1.1 (1996): 7–28.

———. "The Uses of Environmental History." *Environmental History Review* 17.3 (1993): 1–22.

Deloria, Philip J. *Playing Indian*. New Haven: Yale University Press, 1997.

Dunlap, Thomas R. "Aldo Leopold, Wildlife, and the Land Ethic." *Transactions of the Sixtieth North American Wildlife and Natural Resources Conference*. Ed. Kelly G. Wadsworth and Richard E. McCabe. Washington: Wildlife Management Institute, 1995. 521–26.

———. *Saving America's Wildlife*. Princeton: Princeton University Press, 1988.

Egan, Timothy. "Trees Return to St. Helens, But Do They Make a Forest?" *New York Times* 26 June 1988: A1+.

———. "With Fate of the Forests at Stake, Power Saws and Arguments Echo." *New York Times* 20 Mar. 1989: A1+.

Evans, Nicholas. *The Loop*. New York: Dell, 1998.

"Four Gray Wolves Released in Idaho." *New York Times* 15 Jan. 1995: A19.

Gottlieb, Robert. *Forcing the Spring: The Transformation of the American Environmental Movement*. Washington, D.C.: Island, 1993.

"Gray Wolves Await Release in the Rockies." *New York Times* 14 Jan. 1995: A9.

Hays, Samuel P. *Beauty, Health, and Permanence: Environmental Politics in the United States, 1955–1985*. New York: Cambridge University Press, 1987.

———. *Conservation and the Gospel of Efficiency: The Progressive Conservation Movement, 1890–1920*. Cambridge, Mass.: Harvard University Press, 1959.

Isenberg, Andrew C. *The Destruction of the Bison: An Environmental History, 1750–1920*. New York: Cambridge University Press, 2000.

Jacoby, Karl. *Crimes against Nature: Squatters, Poachers, Thieves, and the Hidden History of American Conservation*. Berkeley and Los Angeles: University of California Press, 2001.

Johnson, Dirk. "Yellowstone Will Shelter Wolves Again." *New York Times* 17 June 1994: A12.

Keller, Betty. *Black Wolf: The Life of Ernest Thompson Seton*. Vancouver: Douglas, 1984.

Klein, David R. "The Introduction, Increase, and Crash of Reindeer on St. Matthew Island." *Journal of Wildlife Management* 32.2 (1968): 350–67.

Lears, T. J. Jackson. *No Place of Grace: Antimodernism and the Transformation of American Culture, 1880–1920*. Chicago: University of Chicago Press, 1981.

Leopold, A. S., L. K. Sowls, and D. K. Spencer. "A Survey of Over-populated Deer Ranges in the United States." *Journal of Wildlife Management* 11.2 (1947): 162–77.

Leopold, Aldo. *A Sand County Almanac, and Sketches Here and There*. New York: Oxford University Press, 1949.

Line, Les. "The Endangered Timber Wolf Makes a Surprising Comeback." *New York Times* 26 Dec. 1995: C4.

Lopez, Barry Holstun. *Of Wolves and Men*. New York: Scribner's, 1978.

Lutts, Ralph H. *Nature Fakers: The Romanticizing of Nature*. Golden, Colo.: Fulcrum, 1990.

Matthiessen, Peter. "Our National Parks: The Case for Burning." *New York Times Magazine* 11 Dec. 1988: 38+.

McIntyre, Rick. *War against the Wolf: America's Campaign to Exterminate the Wolf*. Stillwater, Minn.: Voyageur, 1995.

McNamee, Thomas. *The Return of the Wolf to Yellowstone*. New York: Holt, 1997.

Muir, John. *The Wilderness World of John Muir.* Ed. Edwin Way Teale. New York: Houghton, 1975.

Nash, Roderick. *Wilderness and the American Mind.* 3rd ed. New Haven: Yale University Press, 1982.

Peterson, Rolf O. *The Wolves of Isle Royale: A Broken Balance.* Minocqua, Wisc.: Willow Creek, 1995.

"Protecting Wildlife from Man: Drive Is Stepped Up across U.S." *U.S. News and World Report* 25 Nov. 1974: 63–64.

Robbins, Jim. "In Two Years, Wolves Reshaped Yellowstone." *New York Times* 30 Dec. 1997: F1+.

Robinson, William L. "Workshop: Public Relations and Public Education." *The Behavior and Ecology of Wolves.* Ed. Erich Klinghammer. New York: Garland, 1979. 478–81.

Rothman, Hal K. *The Greening of a Nation? Environmentalism in the United States since 1945.* Fort Worth: Harcourt, 1998.

Sale, Kirkpatrick. *The Green Revolution: The American Environmental Movement, 1962–1992.* New York: Hill and Wang, 1993.

Schmitt, Peter J. *Back to Nature: The Arcadian Myth in Urban America.* Baltimore: Johns Hopkins University Press, 1990.

Scott, James C. *The Moral Economy of the Peasant: Rebellion and Subsistence in Southeast Asia.* New Haven: Yale University Press, 1976.

Seton, Ernest Thompson. *Animal Heroes.* New York: Scribner's, 1901.

———. *The Biography of a Grizzly.* New York: Century, 1899.

———. *Lives of the Hunted.* New York: Scribner's, 1901.

———. *Wild Animals at Home.* Garden City: Doubleday, 1917.

———. *Wild Animals I Have Known.* New York: Scribner's, 1898.

Shabecoff, Philip. *A Fierce Green Fire: The American Environmental Movement.* New York: Hill and Wang, 1993.

Slaughter, Thomas. *The Whiskey Rebellion: Frontier Epilogue to the American Revolution.* New York: Oxford University Press, 1986.

Slotkin, Richard. *Gunfighter Nation: The Myth of the Frontier in Twentieth-Century America.* New York: Harper, 1992.

Smith, Henry Nash. *Virgin Land: The American West as Symbol and Myth.* New York: Vintage, 1950.

Speck, Frank. "The Family Hunting Band as the Basis of Algonkian Social Organization." *American Anthropologist* 17.2 (1915): 289–305.

———. "Mistassini Hunting Territories in the Labrador Peninsula." *American Anthropologist* 25.4 (1923): 452–71.

Stevens, William K. "As the Wolf Turns: A Saga of Yellowstone." *New York Times* 1 July 1997: C1+.

———. "Triumph and Loss as Wolves Return to Yellowstone." *New York Times* 12 Sept. 1995: C1+.

———. "Wolf's Howl Heralds Change for Old Haunts." *New York Times* 31 Jan. 1995: C1+.

———. "Wolves May Reintroduce Themselves to East." *New York Times* 4 Mar. 1997: C1+.

Taylor, Alan. *Liberty Men and Great Proprietors: The Revolutionary Settlement on the Maine Frontier, 1760–1820.* Chapel Hill: University of North Carolina Press, 1990.

Thiel, Richard P. *The Timber Wolf in Wisconsin: The Death and Life of a Majestic Predator.* Madison: University of Wisconsin Press, 1993.

Thompson, E. P. "The Moral Economy of the English Crowd in the Eighteenth Century." *Past and Present* no. 50 (Feb. 1971): 76–136.

———. *Whigs and Hunters: The Origin of the Black Act.* New York: Pantheon, 1975.

Trachtenberg, Alan. Introduction to *Ragged Dick; Or, Street Life in New York with the Boot Blacks.* Horatio Alger. New York: Signet, 1990.

Turner, James. *Reckoning with the Beast: Animals, Pain, and Humanity in the Victorian Mind.* Baltimore: Johns Hopkins University Press, 1980.

United States Statutes at Large, v. 86, 92nd Cong., 2nd Sess. Washington: Government Printing Office, 1973.

United States Statutes at Large, v. 87, 93rd Cong., 1st Sess. Washington: Government Printing Office, 1974.

"Vanishing Wildlife." *Time* 8 June 1970: 52–53.

Webb, Walter Prescott. *The Great Plains.* Boston: Ginn, 1931.

Wiese, Thomas F., et al. *An Experimental Translocation of the Eastern Timber Wolf.* Marquette, Michigan: Audubon Conservation Report #5, 1975.

"Wolves Arrive in Wyoming as Battle over Them Goes On." *New York Times* 13 Jan. 1995: A17.

"Wolves in Yellowstone Are Free to Roam after 50-Year Absence." *New York Times* 22 Mar. 1995: A14.

"Wolves Leave Pens at Yellowstone and Appear to Celebrate." *New York Times* 27 Mar. 1995: A8.

Worster, Donald. *Dust Bowl: The Southern Plains in the 1930s.* New York: Oxford University Press, 1979.

———. *Nature's Economy: A History of Ecological Ideas.* 2nd ed. New York: Cambridge University Press, 1994.

Young, Stanley P., and Edward Goldman. *The Wolves of North America.* Washington: American Wildlife Institute, 1944.

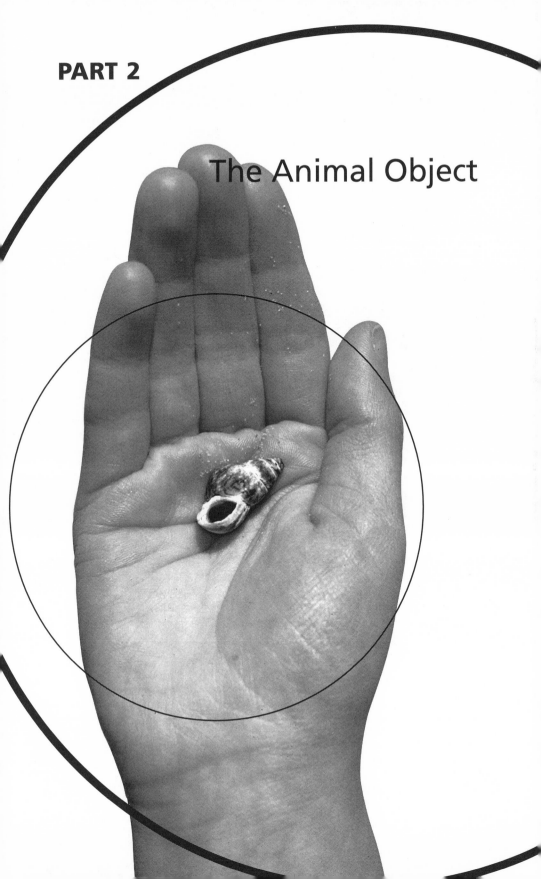

PART 2

The Animal Object

5 What Does Becoming-Animal Look Like?

Steve Baker

Content is easy . . . But form, that's another animal.

—Dennis Oppenheim

An astonishing aside in a particularly dense passage of Deleuze and Guattari's elaboration of the concept of "becoming-animal" in *A Thousand Plateaus* reads as follows: "either stop writing, or write like a rat" (240). This, it seems, may be read as their challenge to the writer or, more generally, to the artist. The artist— the "sorcerer"—has a very particular relation to the animal: "If the writer is a sorcerer, it is because writing is a becoming, writing is traversed by strange becomings that are not becomings-writer, but becomings-rat, becomings-insect, becomings-wolf, etc. . . . Writers are sorcerers because they experience the animal as the only population before which they are responsible in principle" (240).[1] The artist is ultimately responsible to the animal—but what kind of responsibility is this, and to what kind of animal?

Answers are not immediately forthcoming, because the passage moves on to address what the artist is alert to in the animal: "the effectuation of a power of the pack that throws the self into upheaval and makes it reel [*qui soulève et fait vaciller le moi*]. Who has not known the violence of these animal sequences, which uproot one from humanity, if only for an instant, making one scrape at one's bread like a rodent or giving one the yellow eyes of a feline? A fearsome involution calling us toward unheard-of becomings" (240).[2] The themes so fleetingly touched on here—the relation of the human self and the animal pack, the idea that becoming-animal may be something typically experienced "only for an instant," and the implication that the artist's responsibility may be to work fearlessly to prolong such instants—set the broad agenda for the present essay.

In the critical examination of questions of identity which runs through so much poststructuralist and postmodern thought, Deleuze and Guattari's concept of becoming-animal [*devenir-animal*] holds a special place in tying any creative reimagining of the human so closely to that of the animal. The real radicalism of the concept lies not in its reframing of the question of living subjects and their identities, but rather in its charting the possibilities for experiencing an uncompromising sweeping-away of identities, whether human or animal. The individuated subject, the subject of an identity, is a "well-formed" subject (253) which has submitted to those Oedipal forces of fixity, conservatism, and compliance which Deleuze and Guattari so consistently oppose.

Becoming-animal, as they note in their earlier book, *Kafka,* is the creative and "experimental" alternative to this, where "all forms come undone" (7, 13). In their view it is subjects which have forms, and if there is one thing that becoming-animal works against it is the whole "anthropocentric entourage" of the individuated subject (36). Becoming-animal is a means of undoing identity: "There is no longer man or animal, since each deterritorializes the other" (22). Form can only inhibit this radical agenda: "As long as there is form, there is still reterritorialization" (6).

This essay will be concerned with a preliminary interrogation of these seductive but elusive ideas. Why, for Deleuze and Guattari and for others, is the very idea of the animal aligned in some way with creativity? What does it take to gesture toward the other-than-human, and thus to enter that privileged "experimental" state of identity-suspension which they call becoming-animal? If the work of doing so is in some respects the work of the artist, including the visual artist (but that work proscribes the use of identity-tainted form), this raises perhaps the most perplexing question of all: what does becoming-animal look like? Overall, the question is whether or not becoming-animal amounts to something that might be acted on: a practice, in other words, rather than a mere rhetoric.

The answers suggested here will draw on the work and ideas of contemporary artists who use animal imagery, as well as on those of Deleuze and Guattari, looking not only for correspondences among them but for ways in which each might test or illuminate the other. It is therefore likely that such answers will say as much about contemporary attitudes toward art as they do about contemporary attitudes toward the animal—though this may turn out in certain respects to be a fragile distinction.

Art's Becomings-Animal

The videos and video performances made in the 1990s by the British artist Edwina Ashton, such as *Slug Circus* (a living slug and snail moving along and falling off a miniature tightrope, filmed inside a purpose-made marquee), *Sheep* and *Frog* (both featuring the artist herself, uncomfortably dressed in homemade animal costumes), and the bizarre *Bear-Faced Monologue* (fig. 5.1) (in which she animates and gives voice to a slice of cooked meat), are among the very few recent works to be made by an artist with a conscious interest in the philosophical concept of becoming-animal and a positive enthusiasm for Deleuze and Guattari's "hyper-exciting" writing.[3] Nevertheless, becomings-animal are increasingly evident in contemporary art. The last quarter of the twentieth century saw artists' traditional use of animals as little more than remote ciphers for human meanings begin to give way to instances of artist and

68 *Steve Baker*

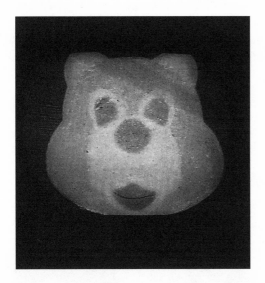

5.1. Edwina Ashton, video still from *Bear-Faced Monologue,* 1996.
Courtesy of the artist.

animal coming closer together as living beings caught up in each other's affairs, willingly or otherwise.

Joseph Beuys's 1974 performance, *Coyote,* in which he and a live coyote improvised their way through their week-long confinement in the Galerie René Block in New York, separated from the spectators by a chainlink barrier, is only the most obvious example of this shift. Its contrivance has become increasingly apparent over the years, and Andrea Phillips has recently doubted the credibility of Beuys's own view of it as "an ecology," suggesting that the artist's position in relation to the animal "is more cruelly focused than that, lying somewhere between ringmaster and slave to the wild coyote with whom he performs" (128).

Like *Coyote,* however, the most telling recent examples of art's becomings-animal do generally involve the pressingly real interaction of artist and living animal. Olly and Suzi (themselves great admirers of Beuys as "the foremost environmentalist in the art world") are British artists best known for painting endangered predators in their natural habitat at the closest possible quarters—whether they be tarantulas and green anacondas in Venezuela, wild dogs in Tanzania (fig. 5.2), or white sharks underwater off the coast of Capetown (fig. 5.3). The two of them work simultaneously on each painting, "hand over hand," as they put it, in conditions that can be both inhospitable and dangerous. Of their work with "polar bears, white sharks, and big cats," for example, they laconically note that "tracking, painting, and interacting with these animals emphasizes the reality of being in the 'food chain.'"

Experiencing these hostile habitats from within is central to the authenticity

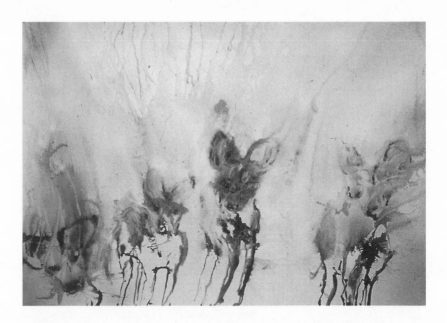

5.2. Olly and Suzi, *Wild Dogs, Tanzania,* 1995. Acrylic and sepia on paper.
© Growbag.

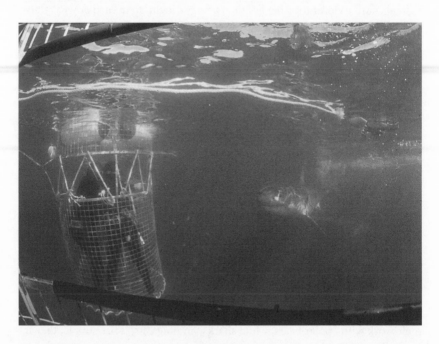

5.3. Olly and Suzi with Greg Williams: the artists at work in a shark cage,
Dyer Island, Cape Town, South Africa, 1997.
© Growbag.

of what they can report both in environmental and in aesthetic terms: "It gives us an immediate response, and that's really what we're after in our work, so the experience comes in our eye and out of our hand. And you can only really get that on site." Acknowledging that generally "the benefits of the interaction are, in the short term at least, in our favor, despite our long-term objectives of helping the animals' predicament," they nevertheless have only limited control of their encounters with these animals. As Olly said to a press reporter about the shark experience, "The visibility was not that great at first, so you see this shape coming out of the gloom and suddenly there is the shark. It is hard to breathe when that happens, let alone paint" (qtd. in Morton 37).

Suspended in the flimsy shark cage, the artists were painting the sharks (with graphite and oil sticks on thick paper mounted on polystyrene) in cramped conditions which were made yet more difficult by the fact that the smaller sharks, driven by curiosity, were easily able to push their heads through the bars of the cage. The photographer Greg Williams, who travels with the artists and documents their animal encounters, was equally uncomfortable: in order to photograph the paintings as they were being made, he had to reach across from his own shark cage to grab theirs with his left hand while holding the camera in his right. Olly and Suzi expected to be afraid (as indeed initially they were), but found themselves after a time more conscious of the "perfect forms" of these creatures when seen in their own environment; their complex aim was to record something of "the wonder, the horror" of the encounter.

Seemingly as far removed from such work as it could be, the filmmaker and performance artist Carolee Schneemann's work with her cats Cluny II in the 1980s and Vesper in the 1990s in fact displays a similarly audacious approach to extreme proximity to the living animal. These are recorded in different versions of a composite photographic piece called *Infinity Kisses* (figs. 5.4 and 5.5), which document the way in which each of these cats, in turn, contributed to and extended the artist's erotic life. Linda Weintraub summarizes Schneemann's daily ritual with Cluny II, who would climb into bed with the artist and her (human) partner: "As she awakened each morning she groped for a camera and aimed it to capture his ardent advances. Close and unfocused, these images capture Schneemann as the blissful recipient of Cluny's deep mouth kisses, his paw embraces encircling her head, and his licking explorations of her lips and neck" (131). Remarkably, her next cat, Vesper, displayed a similar taste for this unusual intimacy, kissing with the artist for anything up to fifteen minutes each morning.

These actions have been described by one (sympathetic) writer as "art bestiality," but the term may be unfortunate insofar as it adds to the ease with which Schneemann's work and its significance—as a recent reviewer noted—are open to being "frightfully misunderstood." As with Olly and Suzi's shark work, there is little apparent interest in shocking the viewer, no matter how disarming the resulting imagery may be. As Weintraub notes, "Schneemann's interspecies eroticism is not exclusively a carnal experience, nor merely a confrontation with society's proscriptions against such sensual behaviors" (130).

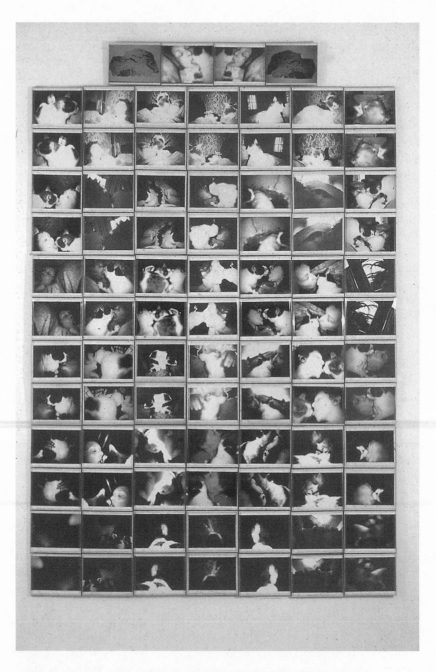

5.4. Carolee Schneemann, *Infinity Kisses*, 1981–88. Wall installation, self-shot
35mm photographs, xerachrome on linen, 140 images.
Courtesy of the artist.

5.5. Carolee Schneemann, *Infinity Kisses,* details.
Courtesy of the artist.

The inescapable sensuality of the pieces is in fact intrinsic to Schneemann's operation as an artist, and it is specifically in this context that she has stated, "I had never accepted that any part of the body be subject to visual or tactile taboos" (qtd. in "Evening"). The cat's input is nevertheless crucial: "pets can teach us pleasure and shamelessness," she remarks (qtd. in Weintraub 129–30). More than this, her cats are explicitly acknowledged to have their own perspective, their own vision of the world, from which the artist can learn something as an artist. In a short text called "Animal," she writes of the lasting influence of her first cat, Kitch, on her many subsequent animal pieces: "Her steady focus enabled me to consider her regard as an aperture in motion."[4] Here is perhaps the perfect image of the work of the artist's becoming-animal: to become, like Kitch, the artist's means of seeing, an aperture in motion, prolonging what Deleuze and Guattari call the "violence" of those instants of upheaval for the human self in which it may take on, for example, "the yellow eyes of a feline" (*Thousand Plateaus* 240). Write like a rat; see like a cat?

Any lingering doubts as to the seriousness and non-sensationalism of Schneemann's being-bound-up-with the animal in her work are dispelled by the remarkable multi-channel video installation *Vespers Pool* (2000). This is art as mourning: as profound and moving an act of commemoration of her dead cat Vesper as could be imagined for any human. The video footage includes snippets of the artist and cat kissing, but also of the cat's ordinary daily movements around house and garden: imagery entirely familiar to any fond pet owner. Over this, the multi-layered soundtrack includes an uncomfortable and insistent scraping sound. As this slowly becomes louder and clearer in the mix, it may dawn on the viewer (without the piece ever needing to confirm it) that this is the sound of the spade with which Vesper's grave is being dug. This is art whose proper and serious purpose, as Schneemann herself observed in an early performance notebook, is to "grind in on the senses" (56).

Swept Up

In their intensity, immediacy, intimacy, and urgency, these kinds of art practice might exemplify what Deleuze and Guattari had in mind when they made this manifesto-like statement in *A Thousand Plateaus:* "We believe in the existence of very special becomings-animal traversing human beings and sweeping them away, affecting the animal no less than the human" (237).

So what is becoming-animal? The question is not so much what it is as what it does. For Deleuze and Guattari, what becoming-animal does is close to what art does. In becoming-animal, certain things happen to the human: the "reality" of these becomings-animal resides "in that which suddenly sweeps us up and makes us become" (279). Of any becoming, they write, "We can be thrown into a becoming by anything at all, by the most unexpected, most insignificant of things"—"a little detail that starts to swell and carries you off" (292).

This being swept up, swept away, suddenly, unexpectedly, with which the human nevertheless goes along as if willingly, resembles some of what Deleuze and Guattari say about art. Despite their warnings that "[a]rt is a false concept, a solely nominal concept" (300–301), and that "art is never an end in itself" (187), they see it as having its own politicized work to do:

> [I]t is only a tool for blazing life lines, in other words, all of those real becomings that are not only produced *in* art, and all of those active escapes that do not consist in fleeing *into* art, taking refuge in art, and all of those positive deterritorializations that never reterritorialize on art, but instead sweep it away with them toward the realms of the asignifying, asubjective. (187)

Art's work—moving the human away from anthropocentric meaning and subjective identity—is presented as much the same thing as the animal's work. It is the work of figuring out how to operate other-than-in-identity. The various arts "have no other aim" than to "unleash" becomings (272). Art, it seems, consists in letting fearsome things fly.

This is very much how, in "Coming to Writing," Hélène Cixous also describes the exhilarating and self-consciousness-erasing force, or "breath," which unexpectedly launches the writer into writing:

Because it was so strong and furious, I loved and feared this breath. To be lifted up one morning, snatched off the ground, swung in the air. To be taken by surprise. To find in myself the possibility of the unexpected. To fall asleep a mouse and wake up an eagle! What delight! What terror. And I had nothing to do with it. (10–11)

In contexts such as these the experience of bodies and the experience of art are not easily distinguished.

Breath

One of the more surprising ways in which a connection might be made between art, the artist, and the animal is via this motif of breath, in examples ranging from Cixous's metaphor for creativity to the breath literally shared between Schneemann and her feline collaborators, or to Olly's comment about his difficulty in breathing, let alone painting, as the shark loomed up toward him. Deleuze and Guattari make no noticeable use of the motif themselves, but it does enable certain associations to be made between their work and that of other artists and writers in terms of how to characterize what it is that art does.

A 1993 photograph by the Mexican artist Gabriel Orozco is called *Breath on Piano*. A little out of focus, the photograph still clearly shows a pool of breath momentarily clouding a piano's gleaming polished surface. It seems to be not so much a reference to the life-giving breath of the artist-god as a demonstration of the simplicity with which everyday surroundings can be imaginatively transformed and individuated, moment by moment. Anyone can do this, and that seems to be the point: perhaps everyone should be doing it. The piece has nothing to do with animals, but it is *warm* work, work about the touch of things, work in the world of the living.

Two pieces by Sutee Kunavichayanont, displayed in the 1999 exhibition *Trace*, extend this inclusive idea directly to animal imagery. *The White Elephant* (1999) and the previous year's *The Myth from Rice Field (Breath Donation)* (fig. 5.6) took the form of lifelike and life-sized inflatable latex animals, displayed alongside each other in a largely deflated state on the gallery floor. The elephant lay on its side; the water buffalo hung from the wall by a metal ring through its nostrils. In the catalog Anthony Bond described these loose skins as "reminiscent of hides one might find in an abattoir," but exhibition viewers were invited to inflate them laboriously with their own breath through a series of tubes extending from each body's sagging form, "so that the breath of the human viewer brings the animal to life" (79).

Breath takes on different and more desperate connotations in Dennis Oppenheim's 1989 sculpture *Above the Wall of Electrocution* (fig. 5.7), in which the deformed shapes of a donkey, a pig, a wolf, and two dogs are hung on hooks from an oval steel track. Their heads are plastic masks, each of which grips the polyester carcass of its own body in its jaw. In this animated sculpture, the animals are shown "literally vomiting themselves out," as Jari-Pekka Vanhala puts it:

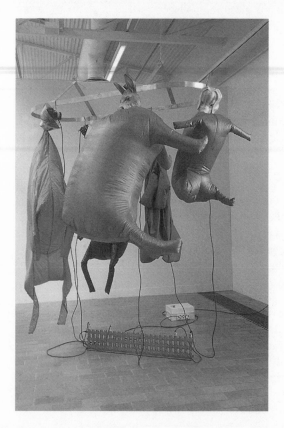

5.6. Sutee Kunavichay-
anont, *The Myth from Rice
Field (Breath Donation),*
1998.
Liverpool Biennial of Contemporary Art, 1999.
Photo: Nick Hunt / JKA.

5.7. Dennis Oppenheim,
*Above the Wall of Electro-
cution,* 1989. Installation
(steel track, cable, animal
masks, electric blowers,
fabric, electric cord, electric plugs, time-delay
relay).
Collection of the Pori Art
Museum, Finland. Photo:
Erkki Valli-Jaakola / Pori
Art Museum.

"The whole body of the animal has become one huge lung, which slowly fills itself until it is trembling with an overdose of oxygen, and then emptying involuntarily leaving behind just a formless pouch. . . . They are doomed to give a desperate kiss of life to themselves again and again" (111–12). Perhaps it is the fact that this is a closed system of vicious and melancholic repetition which makes it seem such a travesty both of breath's and of the animal's association with creativity and life. There is no scope for exchange here, for one body to breathe life into another.

Even (or perhaps especially) when breathing connotes difficulty, its associations tend to be positive and humane. In Jo Shapcott's "The Mad Cow Tries to Write the Good Poem"—one of a series of poems prompted by the epidemic of bovine spongiform encephalopathy (so-called "mad cow disease") in cattle in Britain, which first came to public attention in the late 1980s—it is "in lovely people's ears, their breath, your breath" that the staggering, dying cow hears this elusive poem written (76). The contrast of the Oppenheim and Kunavichay-anont sculptures discussed above is, similarly, of animals effortlessly and (by implication) cruelly inflated by machine on the one hand, and animals effortfully, generously, and never wholly successfully inflated by human breath on the other. As both an embodied thing and a thing capable of moving between bodies with no loss of vitality, breath is perhaps an ideal figure with which to begin to consider what a body is.

What a Body Is

The answer which Deleuze and Guattari seem to offer in *A Thousand Plateaus* to the question of how to operate other-than-in-identity—and of how to operate as an artist—has to do with speeds. To "make your body a beam of light moving at ever-increasing speed," they write, is something which "requires all the resources of art, and art of the highest kind"—the kind of art, that is to say, through which "you become animal" (187).

This making-something of the body involves a radical shift in perspective: a focus, once again, on what things do rather than on what they are. "We know nothing about a body until we know what it can do, in other words, what its affects are" (257). The translator's notes in *A Thousand Plateaus* explain that the word "affect" does not denote a personal feeling [*sentiment*] but rather "a prepersonal intensity corresponding to the passage from one experiential state of the body to another and implying an augmentation or diminution in that body's capacity to act" (xvi). The authors make the implications of this shift in perspective explicit: "Human tenderness" is necessarily foreign to that which "has only affects" (244). Their disdain for a culture locked into its individualistic and possessive concern with sentiments and emotions is nowhere clearer than in their outspoken comments on pets: "individuated animals, family pets, sentimental, Oedipal animals each with its own petty history, 'my' cat, 'my' dog. These animals invite us to regress . . . and they are the only kind of animal psychoanalysis understands" (240). (The Oedipal, for Deleuze and Guattari, represents all that is most timid, petty, conformist, and conservative in human understanding.) Oedipalized beings, whether human or animal, seem to offer no useful model for writing like a rat.

As an alternative to this, the authors seek to describe what they call "a natural play of haecceities, degrees, intensities, events and accidents that compose individuations totally different from those of the well-formed subjects that receive them" (253). "Haecceity" is the crucial term here, the word itself suggesting no more than a kind of "thisness" by which to designate "a mode of individuation" which consists of "relations of movement and rest" and "capacities to affect and be affected." Their examples of haecceities are suitably unbodylike: "A degree of heat, an intensity of white, are perfect individualities" (261). An output of breath might easily be added to such a list.

Individuations experienced as "speeds and affects" are of a quite different order than those of "forms, substances and subjects." A haecceity is not a subject but an event. It involves moving, by whatever means and with the aid of whatever animals, from one perspective to the other: "It is the wolf itself, and the horse, and the child, that cease to be subjects to become events" (262). This ceasing-and-becoming is a matter of desire: "Starting from the form one has, the subject one is," Deleuze and Guattari describe it as the establishing of "the relations of movement and rest, speed and slowness that are *closest* to what one is becoming. . . . This is the sense in which becoming is the process of desire" (272).

Like a work of art, this alternative to identity and to identity-thinking is a willed thing, a worked-on thing: "You are . . . a set of nonsubjectified affects," they say. "You have the individuality of . . . a climate, a wind, a fog, a swarm, a pack. . . . Or at least you can have it, you can reach it." It involves an imaginative rethinking of the body, an inhabiting of that body as haecceity rather than as well-formed subject: "[Y]ou will yield nothing to haecceities unless you realize that that is what you are" (262). Art's rethinking of bodies and forms therefore necessarily includes a rethinking of the artist's own body, understanding it to be distinguished (and experienced) "solely by movement and rest, slowness and speed" (254).

This rhetoric has in fact been readily adopted by contemporary artists and writers (regardless of their limited familiarity with Deleuze and Guattari) to describe their own becomings, whether or not these are explicitly becomings-animal. What is at stake here, to express it rather awkwardly, is a form of inhabiting which is also a kind of unselfing. Martin Amis's aptly named character, John Self, describes the postmodern perception of self-as-event in these terms: "Sometimes I feel that life is passing me by, not slowly either, but with ropes of steam and spark-spattered wheels and a hoarse roar of power or terror. It's passing, yet I'm the one who is doing all the moving. I'm not the station, I'm not the stop: I'm the train. I'm the train" (110).

Schneemann describes her performance practice in rather similar terms: "I'm not ever imagining myself . . . it's the gesture that is inhabiting me. How to inhabit the energy, how fast? How slow?" (qtd. in Eyler n.p.). And Oppenheim, describing some of his more problematic constructions, notes that "in their in-ability to take off, they are nevertheless building up steam" (qtd. in Heiss 179). Here the affects of (and on) artist and artwork are not easily distinguished; Vanhala notes of Oppenheim's art practice that the "process" of making art "is a way to produce a certain experimental condition for an object, body or mind" (112).

Art is a matter of operating at speeds, operating in movement. Ashton writes of her own practice, "If making small films seems no more than being a hamster in its wheel then it's because wheeling is more exciting than not wheeling (a privilege and an indulgence in fact)." And back in the 1960s Schneemann wrote, "Notice this insistence on Motion . . . you can't tell an idea STOP" (50, 60). The animal resonances of such remarks are drawn out in Shapcott's poem "The Mad Cow Talks Back," which begins,

> There are wonderful holes in my brain
> through which ideas from outside can travel
> at top speed . . .

It is about the openness of art, the openness of coming-to-see-as-animal. The cow continues,

> . . . Most brains are too
> compressed. You need this spongy
> generosity to let the others in. (69)

Cixous picks up the theme of openness in her fictional text *Vivre l'orange*, but expresses it in terms of slowness rather than speed: "We do not know how to wait for the other: our waiting attacks. We are pressed, we press and things flee. . . . Things of beauty come to us only by surprise. To please us. Twice as beautiful for surprising us, for being surprised" (108, 110). This surprise, this "other," can of course take animal form: on Ellesmere Island, in the Arctic Cir-cle, it was only after Olly and Suzi had camped out for three weeks that the seldom-seen white wolves which they had gone out there to paint eventually turned up to see what the artists were doing (fig. 5.8).

Vivre l'orange also includes the observation "One must have learned how to inhabit time humanly: to know how to act as slowly, to breathe as deeply as is necessary for a life to grow and think itself humanly. One must be able to live according to the slow seasons of a thought" (76). But as a remark by Schnee-mann suggests, this may just as easily (and just as creatively) be described as learning to inhabit time animally: "My cats teach me how to take time, what to pay attention to," she insists (qtd. in Weintraub 133).

As these remarks also suggest when taken collectively, speed and slowness need not be seen as opposites here. They can both be means of unselfing, means of reaching—as the opening page of *A Thousand Plateaus* puts it—"the point where it is no longer of any importance whether one says I. We are no longer ourselves" (3). This is becoming-animal: "when you become-dog . . . There is no longer a Self [*Moi*] that feels, acts, and recalls; there is 'a glowing fog, a dark

5.8. Olly and Suzi with Greg Williams: the artists painting a white wolf,
Ellesmere Island, 1999.
© Growbag.

yellow mist' that has affects and experiences movements, speeds" (162). And this
unselfing is precisely the work of the artist and writer. In *Kafka,* Deleuze and
Guattari specifically propose that a writer is "an experimental man (who thereby
ceases to be a man in order to become an ape or a beetle, or a dog, or mouse, a
becoming-animal, a becoming-inhuman . . .)" (7). The artist and the animal
are, it seems, intimately bound up with each other in the unthinking or undoing
of the conventionally human.

Fear

What is the relation between fear and this unselfing? The British psy-
choanalyst Adam Phillips addresses something very much like the Oedipal self
when he writes that "the defensive ego has a kind of pre-emptive morality born
of fear, it prejudges in order not to judge, in order not to have to think too
much." Discussing a story concerning fear of a tiger, he also notes that fear "tells
us very little about its object" (47). More than this, fear can produce a kind of
wrong knowledge of its object, its animal, making the sufferer "misleadingly
knowing," so that in this context knowing "becomes rather literally the process
of jumping to conclusions" (59).

80 *Steve Baker*

5.9. Damien Hirst, *Some Comfort Gained from the Acceptance of the Inherent Lies in Everything,* 1996. Steel, glass, cows, formaldehyde solution. The Saatchi Gallery, London. Photo: Stephen White.

Becoming's movement away from the Oedipal human self certainly therefore suggests a moving away from fear, and toward the animal. Everything that is Nietzschean in Deleuze and Guattari seems to support such a view. Nietzsche himself came close to outlining the obligatory and orthodox perspective for the postmodern artist when he wrote of "we fearless ones": those who, as he puts it, cannot persuade their nose "to give up its prejudice against the proximity of a human being" and who "love nature the less humanly it behaves, and art when it is the artist's escape from man" (342).

Simply to align the animal and the admirably fearless artist in opposition to the mass of ordinary timid humans (and their pitiable pets), however, risks doing a considerable injustice to the complexity of the issues involved. Discussing the butchered animals preserved in formaldehyde in his *Natural History* series (fig. 5.9), Damien Hirst remarked, "I want people to be frightened. . . . Frightened of themselves" (qtd. in Morgan 24). In one sense the observation seems to echo Nietzsche's assertion that "it will always be the mark of nobility that one feels no fear of oneself" (236), but Hirst, more interestingly, envisages the *animal* as the source of human fear of the self.

Other statements by contemporary artists dealing with animal imagery unashamedly stress the significance of fear in their own creative practice, in complex and sometimes unpredictable ways. Olly and Suzi state, "Fear plays a vital role in our art-making process." Jordan Baseman, whose work includes highly disturbing and identity-blurring taxidermic constructions (*Be Your Dog* is the title of one of them), writes, "Fear plays a huge role in my work. I think that it is one of the main things that inhibits and dictates behavior in humans. And it is also, I think, one of our prime motivators." Edwina Ashton says of fear, "It's critical, really, in my work. Using animals is a way of avoiding all sorts of things, for me." She also writes more generally of her "impulse to make things" coming from "an excitement and nervousness about the world."[5]

Such remarks need not be read as evidence of the Oedipal emotionalism of which Deleuze and Guattari are so critical. To the extent that fear has a creative role here, it may be less as an emotion than as an *affect,* as part of a set of practices which allows a body to go on, to do things. Writing of their research on and experience of dangerous animals in the wild, Olly and Suzi significantly conclude, "The knowledge we gather arms us, but fear is still present; a warm glow, keeping us warm."

The complex interaction of knowledge and fear figures in Schneemann's thinking on the animal, too. Writing that "I identify with the animal as a feminine principle. I have always known myself to be an animal among animals," she goes on to identify a range of specific knowledges and skills acquired through an upbringing doing farming work, which developed into a broader politicized awareness of the brutal and cynical decision-making which is imposed on all aspects of the non-human world. Fear therefore figures in her work, at least in part, as something inextricable from an informed rage—a thing with its own quite specific work to do:

> [M]y animal fear goes deep; deep as rape, genital mutilation, witchcraft burnings, religious persecution, because my culture has defined itself for 2000 years (until "yesterday") by the systematic exclusion of participatory feminine powers. I recognize meat fear that is soft, yielding, tender, moist; for the creatures outrun, overwhelmed, trapped, and stripped of their sacral realm.

Sue Coe's position, as is absolutely clear from her work (fig. 5.10), also involves a sympathetic identification with the animal. That identification sees productive, affective fear as something entirely different from emotional timidity: "The difference between humans and non-human animals is that the latter do not live in fear, once the threat has moved on by, they regain joy, and carry on. . . . Humans are not so healthy, every moment has some level of anxiety, aggression and fear."

Devising a means of unselfing, of becoming-animal, through the work— loosely speaking—of writing like a rat, may be one of the few ways of getting

5.10. Sue Coe, *Rhino in Belgrade*, 1999. Lithograph.
Copyright © 1999 Sue Coe. Courtesy Galerie St. Etienne, New York.

to what Coe and Schneemann both call "joy," and what *A Thousand Plateaus* (a little surprisingly) terms the body's potential for "gaiety, ecstasy, and dance" (150).

Involution

Deleuze and Guattari's "write like a rat" paragraph, it will be recalled, closes with a reference to a "fearsome involution calling us toward unheard-of becomings" (240). That word "involution" does complex work in *A Thousand Plateaus*. Clarifying their concept of becoming, in which different beings—human and animal, for example—enter into temporary blocks of becoming, they insist that this is always a relation of alliance rather than of filiation: "There is a block of becoming that snaps up the wasp and the orchid," they write of one of their most famous examples, "but from which no wasp-orchid can ever descend." Their preferred term to describe this relation of alliance is involution, "on the condition that involution is in no way confused with regression" (238).

It therefore seems to be the word's connotations of enfolding or entanglement which attract them. "Becoming is involutionary, involution is creative" [*Le devenir est involutif, l'involution est créatrice*], they declare unambiguously (238). Earlier in the book they describe the imaginatively unformed and rethought body, the "body without organs," in similar terms: "It is an involution, but always a contemporary, creative involution" (164).

When it involves the animal it may be "fearsome" [*terrible*], but this entanglement is nonetheless creative. Why? Perhaps because its work lies in part in the undoing of simplistic binary thought. "Fear as well as joy and rage tangle in the motives in my work," writes Carolee Schneemann, and other artists describe similar entanglements in their dealings both with the animal and with art. Olly and Suzi speak of "the wonder, the horror" of their shark encounters. Of her own creative experience, Cixous declares, "What delight! What terror." Ashton describes her animal pieces as an attempt to reach "a dainty balance simultaneously of extreme joyfulness and soft cynicism (stupidity)." And Coe's account of animals' healthy alternation of "joy and fear," as if "as one," may describe a comparable involution.

Involution, then, might be thought of as part of art's continuing creative work against regression into the individualistic and fear-ridden Oedipal human self. As Deleuze and Guattari put it in *Kafka*, "One allows oneself to be re-Oedipalized . . . by fatigue, by a lack of invention" (33). The perplexing qualities of fear find their own inventive place in this involution. As Adam Phillips notes, "fear, in its very disarray, orientates us" (56).

Form

Fear may be regarded as part of the work undertaken by the artist in respect of the animal, and fear's capacity to orient the artist brings the present discussion back to the difficult subject of form. For all the talk of being "swept up," the fact that art and becoming-animal undertake similar kinds of work means that both involve a degree of conscious, considered decision-making—orientation—in which the question of form is central. As Joseph Beuys recognized, the postmodern cliché that "everything is possible" is of little use to the artist, who must work to arrive at "a precisely worked-out form" (1035). The question to be addressed in relation to form might therefore not simply be "what does becoming-animal look like?" but rather—in the light of Deleuze and Guattari's apparent creative proscription on form—"why do so many of contemporary art's becomings-animal still look like an animal?"

Beuys's reservation about the postmodern preference for "openness" is in fact by no means at odds with Deleuze and Guattari's approach. In *Kafka* they note that becoming-animal is achieved "through a style . . . and certainly through the force of sobriety" (7). A similar point is made in *A Thousand Plateaus*, where they note that becomings demand "much sobriety, much creative involution" (279). In that book's discussion of the "body without organs" (or "BwO"), this sobriety specifically takes the form of "caution" [*la prudence*]. This has nothing to do with Oedipal fearfulness, of course: "Where psychoanalysis says, 'Stop, find your self again,' we should say instead, 'Let's go further still, we haven't found our BwO yet, we haven't sufficiently dismantled our self.' Substitute . . . experimentation for interpretation" (151). At the same time, it has nothing to do with wild abandon: "Were you cautious enough? Not wisdom, caution. In doses. As a rule immanent to experimentation: injections of caution" (150). Caution is itself an "art" (160).

In asking, "How do you make yourself a body without organs?" they warn that "you can botch it" [*vous pouvez le rater*] (149), which already suggests that the form which it takes is a matter of importance. In pursuit of the intensities that mark out the body without organs, the danger is of an uncontrollable falling-apart—of "wildly destratifying" (160) rather than meticulously disarticulating the body. The result can be the "dreary parade" of "emptied bodies" produced in the extremes of masochism or of drug abuse, where the body is reworked in a manner drained of creativity or true experimentation, on its way to becoming "a body of nothingness, pure self-destruction whose only outcome is death" (150, 162).

There are striking connections here with Dennis Oppenheim's comments on how artists might handle the volatility of their experience. In the long interview with Heiss (in which he refers to his limited knowledge of the writings of Derrida and of Baudrillard, but gives no indication that he is aware of Deleuze and Guattari's work, despite the extraordinarily Deleuzian tone of some of his remarks), he proposes that "artists, quite often more than other people, are thrown into volatile storms as they evolve. This has always been what I thought art's content was—a way of feeding this experience into a form" (170). Oppenheim, who frequently characterizes both his self-perceptions and his work in terms close to Deleuze and Guattari's notion of the haecceity, goes on to discuss his openness to what he calls "internal and external weather conditions entering my work":

A period of operation on unstable ground is something I'm very familiar with. The act of making art for me is as if you are falling. The ground moves away, and you hold your breath and fall. . . . My overwhelming sensation in entering work has always been a rapid heart beat. I shake, then a tremor begins. By then I know I'm falling, and I begin to throw out images as if one can suspend me, carry me away. You want to throw out the images that can save your life. . . . You need a coherent image structure before you hit the ground, otherwise you go right through and keep falling. (172)

It is worth returning to some of the detail which Deleuze and Guattari give in *A Thousand Plateaus* about the body without organs. The work, the practice of constructing that body, is already close to the practice of making art, of becoming-animal, of writing like a rat: "People ask, So what is this BwO?—But you're already on it, scurrying like a vermin," they insist (150). In a passage which seems especially revealing about their views on art and its implications for the question of form, they write,

What does it mean to disarticulate, to cease to be an organism? How can we convey how easy it is, and the extent to which we do it every day? And how necessary caution is, the art of dosages, since overdose is a danger. You don't do it with a sledgehammer, you use a very fine file. You invent self-destructions that have nothing to do with the death drive. Dismantling the organism has never meant killing yourself, but rather opening the body to . . . distributions of intensity, and territories and deterritorializations measured with the craft of a surveyor. (159–60)

The quotidian nature of this creative practice is important: it is there in everything which unhumans the human, "making one scrape at one's bread like a rodent" as well as "write like a rat" (240).

Elsewhere in *A Thousand Plateaus,* the dangers that accompany creative becomings are described in less apocalyptic terms which will be recognized by any artist or writer. The purpose of caution is to prevent the work (of the body, or of art) "from bogging down, or veering into the void" [*de s'enliser, ou de tourner au néant*] (251).

Like writing, all such becomings are achieved through style and sobriety. They concern the fragility of art's (and of the artist's) access to the animal. As the animal advocate Linda Vance pointedly observes about any writing on the animal which, taking the easy route, chooses to moralize and to interpret, "The frogs are effectively gone . . . they disappear as soon as I impose a narrative on them." She couches her alternative in the language of art, and of art's work: "So sharpen your carving tools. . . . Clear a space. There will be work to do" (165–66). That work is an unmaking of the secure and fearful self as much as a making of art. It is an art, to borrow Adam Phillips's tantalizing words, in which "the idea of human completeness disappears," and whose difficult purpose is at least in part to offer "good ways of bearing our incompleteness" (7).

This can be a matter of learning how to operate alongside the animal, alongside fear, being open to both. Olly and Suzi, six hours upriver by motor canoe

5.11. Olly and Suzi with Greg Williams: the artists painting tarantulas,
Venezuela, 2000.
© Growbag.

in Venezuela in February 2000 on a trip to paint endangered green anacondas
as well as ants and tarantulas, are interviewed at sunset by a radio reporter who
has traveled with them. The two artists are working, on the jungle floor, on one
of the tarantula paintings (fig. 5.11). (The "V" shape seen in the upper left of
the photograph is that of the two brushes held by the artists.) Asked by the reporter about the difficulties of working in such conditions, Olly—a degree of
fear audible in his breathless response—remarks, "The proximity. There's no
space. We're like a foot and a half away from a spider that's perfectly capable of
jumping on your head from there—which worries me." Suzi, addressing both
the reporter and Olly, describes their improvisatory practice: "We've now swapped
places and we're just moving around the piece, working over each other. . . .
That is *such* a huge spider—it's very beautiful actually. Can you keep that a bit
thinner, that line? Look, she's rearing up—whoa!" Olly continues, "The whole
of our work is about process, and it's about chance. I don't know what Suzi's
going to do, she doesn't know what I'm going to do, we don't know what the
animal's going to do, what the environment's going to do."[6]

Dennis Oppenheim's comments to Heiss, years earlier, about his own "desperate attempt to describe the edges of form" might serve as an interesting if

unwitting commentary on both the openness and the urgency which inform work such as that of Olly and Suzi. "Content is easy," Oppenheim suggested. "But form, that's another animal."

> I guess when you fall, you move your arms in a certain way, trying to save yourself; you're grabbing at air . . . and you shape it out of the desperation. . . . You're breathing in form, and you can tell because things are attaching themselves, going one way and not the other. There's always a shortness of breath; you're panting, as if you're breathing around a box. It's form! Despite yourself, form is the blanket you catch yourself in. Inspired form is pulled out of the atmosphere and made thick by your need to survive. (182–83)

Here breathlessness, fear, and not-knowing all play their part in the cautious construction of the body without organs, the form of the becoming-animal, in which artist, animal, and artwork are swept up into a sometimes uncomfortable and always unpredictable proximity. (Would it be too glib to suggest that art steers through fear via form?) If the process is seen by artists as unquestionably worthwhile, this is perhaps because fear, according to Adam Phillips, itself "signifies proximity to something of value" (56).

Oppenheim's concern "to describe the edges of form" (form which is "made thick by your need to survive"), and Suzi's casual request to Olly—"Can you keep that a bit thinner, that line?"—draw attention to something which is too easily overlooked in written accounts of visual art: the centrality of the mark, and of mark-making. Orozco's *Breath on Piano,* which perhaps shows the artist's mark at its most unadorned, deliberately understates its importance. Wendy Wheeler makes more of it in her speculations on art and "what it means to be human." Making art, she suggests, is a social ritual which "makes a mark which must contain (in both senses) elements of excess and abandonment." Art is thus both "abandonment to affect (becoming animal)" and "a marking off (from the animal) of what it means to be a human animal." In this sense, she proposes, art can indeed be seen as "a way of warding off fear."[7]

It is artists' refusal to be bound by even such humane distinctions as these, however, which allows—if only on rare occasions—the production of an art in which the animal-made mark can also play a vital role. Olly and Suzi nonchalantly acknowledge that "animals are too busy" getting on with their everyday lives "to care too much about art," but wherever it is possible without too much manipulation of the situation, they allow the animals depicted in their work to "interact" with the work and mark it further themselves. This may take the form of prints or urine stains left on an image by a bear or an elephant, or of chunks bitten off by a wolf or a shark, or may simply be the muddy trace of an anaconda that has moved across a painting (fig. 5.12).

Once marked by the animal, these pieces are described by the artists as "a genuine artifact of the event," and are intended above all to bring home the truth and immediacy of these animals' precarious existence to a Western audience which has grown largely indifferent to the question of endangered species. For there to be an animal-made mark, the animal has to be present, and has to

5.12. Olly and Suzi with Greg Williams: *Green Anaconda*, Los Llanos, Venezuela, 2000.
© Growbag.

participate actively (if unwittingly). What is performed through its presence and recorded in its marks is precisely that animal's reality.

In this sense the pieces stand as a direct and urgent challenge not so much to postmodern aesthetic sensibilities as to what Hans Bertens has called "a deeply felt loss of faith in our ability to represent the real" in a postmodern world (11). Like all becomings-animal, whether as experienced or as presented by the artists, such pieces are—if nothing else—the mark, and the result, of an encounter with the real. Despite this, and despite the fact that Deleuze and Guattari might hardly have approved such a description, becoming-animal must nevertheless be understood to be an aesthetic project—albeit a peculiarly awkward one.

Flawed, Unfitting . . .

On the question of form, Deleuze and Guattari's position is not dissimilar to Lyotard's broadly contemporaneous proposal, in *The Postmodern Condition*, that the postmodern is "that which denies itself the solace of good forms" (81). Good form, like Deleuze and Guattari's notion of the "well-formed," here represents the most lamentably timid option: the very antithesis of art. In contrast, the thing which renders so many of the animals in contemporary art post-modern is precisely their botched or problematic relation to form. Titles sometimes make the theme explicit, in examples ranging from Jo Shapcott's poem "The Mad Cow Tries to Write the Good Poem" to Dennis Oppenheim's 1989 sculpture *Badly Tuned Cow*. As Oppenheim remarked to Germano Celant of his work in general, "I am not driven to make a work about the sublime or the beautiful" (47).

This is an aesthetic rooted in conceptions of the artist's own experience as an unfitting human, a becoming-animal. In Will Self's novel *Great Apes*, the central character—the artist Simon Dykes—feels like a stranger in his own body from the start. He thinks of his body as "this physical idiot twin," and finds himself experiencing "the psychic and the physical ever so slightly out of registration." This is what he calls "that lack-of-fit" (11–12). Later in the novel, Dykes wakes one morning (after a night of particularly heavy recreational drug abuse) to find that he has turned into a chimpanzee, as have all other inhabitants of his previously human world. His unwillingness to accept his new animal status is itself described in terms of a physical mismatch, "as if the limbs he were attempting to control were not altogether coextensive with those he actually had" (126), but it could be said that the form of his animal body now at least made sense of his own ill-fitting human experience.

Without overstating the extent to which such discomfort—David Williams enticingly calls it "the feral dramaturgy of fucked-up-and-yet-ness" (137)—may now be characteristic of the compromised beings who inhabit a postmodern world, contemporary art is certainly awash with unfitting animal becomings. Examples range from William Wegman's dog famously got up to look like an elephant or a frog to Ashton's irritable inhabiting of her homemade and "haphazard" animal costumes in her video performances *Sheep* and *Frog*—provisional creatures whose form, though recognizable, never quite coheres.

Jan Fabre's extraordinary video installation at the Natural History Museum in London, early in 2000 (fig. 5.13), illustrated how this postmodern aesthetic may be understood to operate. It involved a handful of the museum's entomologists, wearing goggles and makeshift insect costumes which were clumsily strapped on over their overalls, clanking noisily around the museum's galleries and variously imitating flies, beetles, butterflies, and parasitic wasps as they simultaneously (and inaudibly) held forth on their specialisms. They looked like nothing so much as those clips of early-twentieth-century film, incorporated with such glee into Peter Greenaway's film *The Falls*, which show men with huge wings strapped to their flailing arms making futile attempts at human flight. But in Fabre's visual fable of experts performing inexpertly, it was the scope for exchange and becoming which was central. Called *A Consilience*, his vision was of "a sharing of knowledge," across disciplines and perhaps also across species.

In these circumstances, human fallibility and failure will be ever-present. The British artist John Isaacs describes an early work which, like his better-known *Untitled (Dodo)*, also had "vestigial wings." It was a bicycle with a flying mechanism which allowed him to explore "how possible it is to escape your perspec-

5.13. Jan Fabre, still from the video installation *A Consilience*, 2000.
© The Natural History Museum, London.

tive"—and indeed how difficult it is to do so. This melancholy bike-bird, a deliberate reference to the "futility" of repetition in art ("the moment of inventing flight for humans has passed"), unsurprisingly refused to take off, refused to sweep the artist up into a becoming-animal despite his "furious pedalling." But this of course was the whole point: "I knew it wouldn't work; it was built not to work."

Both in art and in life, according to *A Thousand Plateaus*, a consequence of becoming is that too much human planning "will always fail" (269). (The verb used is again *rater.*) Becomings-animal in art can render this explicit. Like Isaacs's flightless assemblage, Sutee Kunavichayanont's inflatable latex animals (fig. 5.6) are stubbornly unworkable. They demand too much breath—the huge recumbent elephant even more so than the water buffalo—so whole teams of humans are needed to get any visible sign of life into their limp bodies. In this instructive mismatch, it seems that only human multiplicity will enable these singular forms to become-animal.

This is central to Deleuze and Guattari's concerns: "what interests us are modes of expansion. . . . The wolf is not fundamentally a characteristic or a certain number of characteristics: it is a wolfing. The louse is a lousing, and so on. . . . every animal is fundamentally a band, a pack," they write in *A Thousand Plateaus* (239). (As much as an animal, the becoming produced in Kunavichayanont's pieces is therefore a humaning.) In *Kafka*, they propose that to become-animal is "to find a world of pure intensities, where all forms come undone," as do all meanings (13). Their fascination with pack modes and with other forms of animal multiplicity leads them even to assert that individual animals are "still too formed, too significative, too territorialized" (15).

While pack imagery in itself is comparatively unusual in contemporary art (fig. 5.2), the form of the individual, recognizable animal is put under considerable pressure in other ways. Hirst's butchered pieces render the single animal multiple (fig. 5.9), its own becoming-pack, testing the limits of its formal endurance. Ashton's *Bear-Faced Monologue* (fig. 5.1) deals differently with its animal's sliced multiplication. Working with a product called *Billy Bear*, briefly available from a British supermarket, consisting of precooked slices of pork and chicken pressed into the shape of a bear's face (multiple becomings-animal before the artist even got at it), her video animated the product in order to test its conceptual limits. Giving the face a voice ("I love nature") and even a song ("Oh Billy Bear all face no hair / You are the one whose life I share," et cetera), she aimed to push it close to the breaking point by "providing this slice of stupid meat with things it can't support, like opinions, and love affairs."

The uncomfortable reality and flawed form of becoming-animal can equally take the shape of bodies bound together rather than pushed or pulled apart. Throughout her work Sue Coe uncompromisingly presents this idea of humans and animals bound together by their (generally inhuman) circumstances. Her 1999 print *Rhino in Belgrade* (fig. 5.10) carries a text which reads, "Belgrade Zoo 99 / A rhino driven mad by night bombing kills herself by repeatedly smashing her head against the wall." The rage and despair which might be attributed anthropomorphically to the animal itself, or more directly to the artist seeing the animal in these circumstances, is given formal expression here through the human spectators above the rhinoceros, who clasp their downcast heads in their hands. The print's literary equivalent might be something like Lucy Lurie, the daughter of the disgraced professor in J. M. Coetzee's novel *Disgrace*, trying to explain to him the quotidian reality of contemporary life: "there is no higher life," she says. "This is the only life there is. Which we share with animals" (74).

Such acknowledgments of being bound up with each other, inhabiting each others' lives and spaces, might be seen as an aspect of the artist's responsibility toward the animal, to which Deleuze and Guattari allude. Many years earlier, Heidegger addressed something similar when he described serious philosophical attempts to understand animal life in terms of the human "being able to go along with the other being while remaining *other* with respect to it." This he envisaged as a going-along-with undertaken for the sake of "directly learning how it is with this being" (202–203).

Responsibility, of course, may also figure as form. The artist's work, simultaneously cautious and experimental, is to find an appropriate form or style for becoming-animal, for writing like a rat. This in no sense rules out unfitting or flawed or ragged form, which can itself be a proper resistance to the complacency of the "well-formed" subject. And just as "the most unexpected, most insignificant of things" can hurl the human into a becoming-animal, it may be that the same small detail constitutes appropriate form. Deleuze's own attempts to write like a rat, for example, are said to have sometimes included a refusal to cut his nails.[8]

. . . Beautiful?

This leaves the genuinely difficult question of beauty. Sue Coe's work focuses mainly on the inhumanity with which animals are treated, and is therefore in large part an imagery of animal distress, but for other artists who share her direct concern for animal well-being—such as the painters Olly and Suzi and the photographers Frank Noelker (fig. 5.14) and Britta Jaschinski (fig. 5.15) —their perception of animals as "beautiful" is central to their aesthetic concerns. Can beautiful form be distinguished from good form, and redeemed in some way from the bad name which Lyotard has so persuasively given to "good forms" as a source of uncritical (and irresponsible) "solace"?

Clearly it can, even in a medium like photography, which might superficially be thought of as tied to the truthful representation of animal form. For both Jaschinski and Noelker, the photographer's responsibilities lie in the detailed and difficult work of looking, and of communicating critical knowledge in that looking. Jaschinski has little time for wildlife photography which abnegates this responsibility, saying of a picture by one well-known photographer, "Visually this is absolutely beautiful . . . but I'm not sure what this image is actually telling me." Of the majority of such wildlife photographers, she says, "the problem I have is that they just make it far too comfortable for their viewers."

Noelker's comments on beauty are similarly qualified. Although he tries "for the strongest combination of beauty and sadness I can get," he says of the animals he photographs in zoos that "they don't fit" (Noelker n.p.). The difficult work of animal form—precisely the artist's work—is to make that lack of fit evident. Here beauty becomes something of an irrelevance except to the extent that artists are themselves swept up in the events their work initiates. As Deleuze and Guattari put it in *A Thousand Plateaus*, artists "become-animal at the same time as the animal becomes what they willed" (305). These "animal sequences," these becomings-animal, where vacillating form makes the viewing self "reel," affect "the animal no less than the human," and are all about dodging recognizability, eluding identity, "if only for an instant" (237, 240). This is what becoming-animal looks like.

These mundane misrecognitions may sweep up a swimming polar bear (fig. 5.14), for example, into something very like an oversized hand or claw. Their mundanity makes them no less a source of beauty and delight. In one of her poems Jo Shapcott awkwardly adopts the voice of the cartoon cat Tom (from "Tom and Jerry") in order to describe just such effortless visual liberties:

> I can't get this new voice to explain to you
> the ecstasy of the body when you fling

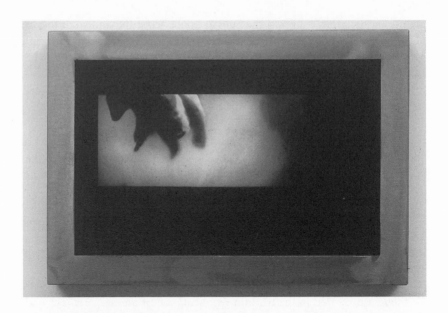

5.14. Frank Noelker, *Polar Bear,* 1994. Detail from a grid of twelve dye subli-
mation images in metal frames.
Courtesy of the artist.

5.15. Britta Jaschinski, *Pongo pygmaeus.* Photograph from the series *Beasts,*
1998.
Courtesy of the artist.

yourself into such mayhem, open yourself
to any shape at all. . . . (39)

It must be stressed that this proliferation, the animal's becoming-pack, is not an abandonment of form. Even Deleuze and Guattari acknowledge that it is "necessary to retain" a certain bodily integrity, to prevent "a regression to the undifferentiated." They describe the process of form's own becoming: "There is a material proliferation that goes hand in hand with a dissolution of form (involution) but is at the same time accompanied by a continuous development of form" (270).

The visual expression of this development is neither possessive nor prescriptive, but seeks instead to test form, to push and to pull at it, working the animal free of its human confines. Jaschinski's remarkable *Beasts* series, completed in the late 1990s, is driven by her conviction that "animals don't need us." It aims to chart the animal's line of flight from the human, and to reinforce the animal's unavailability to the human. The dense black centers of these often unfocused images show the animals keeping knowledge of their bodies to themselves, and refusing to be easily drawn out about what it is that they are (fig. 5.15). Asked how she regards the status of the animals in these images, Jaschinski explains, "Initially the idea was not even representing a species, but just a creature—anything but the human being, basically; *anything*."

One consequence of this radical un-humaning of animals, which attends instead to what she calls "their own existence . . . their dignity and their beauty," is a visual ambiguity which loses all sight of taxonomic propriety. In one photograph, the body of a zebra could almost be mistaken for that of an exotic translucent fish; in another, the ears of a llama have more than a passing resemblance to the wings of a butterfly. Animal unconcern here takes the form of the artist's unconcern: she says of her extraordinary gibbon photograph, *Hylobates lar* (fig. 5.16), that "it really doesn't matter" that it is sometimes mistaken for a frog. The disorientation is deliberate. "That's what I'm interested in," she says; "on first glance you don't even recognize it as an animal." The work of the image lies elsewhere, in sloughing off preconceptions and recognizable identities, and in discouraging anthropomorphic identifications.

Like much of the art discussed in this essay, this work takes the notion of the artist's responsibility to the animal more literally than Deleuze and Guattari may have envisaged. Animals, for Deleuze and Guattari, seem to operate more as a device of writing—albeit a device which initiated its own forms of political practice—than as living beings whose conditions of life were of direct concern to the writers. In a revealing remark in *Kafka*, they note that Kafka's "animal tales" were mainly written before *The Trial*, a novel "which liberates itself from all animal concern to the benefit of a much higher concern" (15). In contrast, Jaschinski and other like-minded artists refuse such hierarchical thought and indeed care little for the human's own desire to become-animal. Their achievement is to have come so close to giving visible form to what is animal in the

5.16. Britta Jaschinski, *Hylobates lar*. Photograph from the series *Beasts*, 1998. Courtesy of the artist.

animal. In doing so they enable the viewer to glimpse and perhaps even to be swept up in something of the animal's difference and distance from the human.

Notes

1. I am indebted to Stephanie Rowe, who drew my attention to Deleuze and Guattari's views on the artist's responsibility to the animal when the paper on which this essay is based was first delivered, at the Representing Animals conference in April 2000.

2. In quoting from *A Thousand Plateaus*, I have occasionally (as here) incorporated brief phrases from the French original, *Mille plateaux*, where this has seemed to clarify or to add to the sense of Brian Massumi's generally excellent translation.

3. Here and elsewhere in this essay, statements by contemporary artists are (unless otherwise indicated) drawn from unpublished interviews, conversations, and correspondence with the author between 1998 and 2000.

4. From "Animal," 2000 (though the section quoted was first written in 1977). I am most grateful to Carolee Schneemann for a copy of this unpublished text.

5. Artists' comments in this and in the following paragraphs are drawn from cor-

respondence and conversations prompted by a question I circulated in July 2000 about the role played by fear in these artists' work. More generally, my discussion of this issue owes much to the thoughtful responses of Stephanie Rowe and Wendy Wheeler on the broader question of the extent to which fear might itself have a creative dimension.

6. Olly and Suzi, interviewed by Huw Williams. The report was broadcast on the BBC World Service program *Outlook* on 17 March 2000.

7. Wendy Wheeler, correspondence with the author, July 2000. For a fuller account of the significance of Wheeler's ideas on art and the animal, see my *The Postmodern Animal*, 16–17 and 164–65.

8. I am grateful to Bülent Diken, who knew Deleuze and Guattari, for this anecdotal detail.

Bibliography

Amis, Martin. *Money: A Suicide Note.* London: Jonathan Cape, 1984.

Baker, Steve. *The Postmodern Animal.* London: Reaktion, 2000.

Bertens, Hans. *The Idea of the Postmodern: A History.* New York: Routledge, 1995.

Beuys, Joseph. "The Cultural-Historical Tragedy of the European Continent." *Art in Theory, 1900–1990: An Anthology of Changing Ideas.* Ed. Charles Harrison and Paul Wood. Oxford: Blackwell, 1992. 1032–36.

Bond, Anthony. "Artists' Pages." *Trace: 1st Liverpool Biennial of International Contemporary Art.* Exhibition catalog by Anthony Bond. Liverpool: Liverpool Biennial of Contemporary Art, 1999. 24–163.

Celant, Germano. "Conversation between Germano Celant and Dennis Oppenheim." *Dennis Oppenheim.* Exhibition catalog by Germano Celant. Milan: Edizioni Charta, 1997. 17–53.

Cixous, Hélène. *"Coming to Writing" and Other Essays.* Ed. Deborah Jenson. Cambridge, Mass.: Harvard University Press, 1991.

———. *Vivre l'orange / To Live the Orange* [adjacent French and English texts], 1979. Reprinted in *L'heure de Clarice Lispector.* Paris: Édition des Femmes, 1989. 7–113.

Coetzee, J. M. *Disgrace.* London: Secker and Warburg, 1999.

Deleuze, Gilles, and Félix Guattari. *Kafka: Toward a Minor Literature.* Trans. Dana Polan. Minneapolis: University of Minnesota Press, 1986.

———. *Mille plateaux: Capitalisme et schizophrénie 2.* Paris: Minuit, 1980.

———. *A Thousand Plateaus: Capitalism and Schizophrenia.* Trans. Brian Massumi. London: Athlone, 1988.

"An Evening with Carolee Schneemann." Publicity handout. Department of Film, Museum of Modern Art, New York. 7 December 1992.

Eyler, Carolyn. "Carolee Schneemann: Drawing Performance." *Carolee Schneemann: Drawing Performance.* Exhibition catalog by Carolyn Eyler. Gorham and Portland: University of Southern Maine Art Galleries, 1999. Unpaginated.

Heidegger, Martin. *The Fundamental Concepts of Metaphysics: World, Finitude, Solitude.* Trans. William McNeill and Nicholas Walker. Bloomington: Indiana University Press, 1995.

Heiss, Alanna. "Another Point of Entry: An Interview with Dennis Oppenheim." *Dennis Oppenheim: Selected Works, 1967–90 / And the Mind Grew Fingers.* Exhibition catalog by Alanna Heiss. New York: Institute for Contemporary Art, P.S.1 Museum, 1992. 137–83.

Lyotard, Jean-François. *The Postmodern Condition: A Report on Knowledge.* Trans. Geoff Bennington and Brian Massumi. Manchester: Manchester University Press, 1984.

Morgan, Stuart. "An Interview with Damien Hirst." *Damien Hirst: No Sense of Absolute Corruption.* Exhibition catalog by Stuart Morgan. New York: Gagosian Gallery, 1996.

Morton, Cole. "Watch Your Back. Mr. Hirst." *Independent on Sunday* (London) 15 March 1998: 36–37.

Nietzsche, Friedrich. *The Gay Science.* Trans. Walter Kaufmann. New York: Vintage, 1974.

Noelker, Frank. "Zoo Pictures." Paper delivered at Millennial Animals, a conference hosted by the Department of English Literature, University of Sheffield. July 2000.

Phillips, Adam. *Terrors and Experts.* London: Faber and Faber, 1995.

Phillips, Andrea. "A Dog's Life." *Performance Research* 5.2 (2000): 125–30.

Schneemann, Carolee. *More Than Meat Joy: Complete Performance Works and Selected Writings.* Ed. Bruce McPherson. New York: Documentext, 1979.

Self, Will. *Great Apes.* London: Bloomsbury, 1997.

Seppälä, Marketta, Jari-Pekka Vanhala, and Linda Weintraub, eds. *Animal. Anima. Animus.* Pori, Finland: Pori Art Museum Publications 43, 1998.

Shapcott, Jo. *Her Book: Poems, 1988–1998.* London: Faber and Faber, 2000.

Vance, Linda. "Beyond Just-So Stories: Narrative, Animals, and Ethics." *Animals and Women: Feminist Theoretical Explorations.* Ed. Carol J. Adams and Josephine Donovan. Durham, N.C.: Duke University Press, 1995. 163–91.

Vanhala, Jari-Pekka. "Dennis Oppenheim: Death and Breath." Seppälä, Vanhala, and Weintraub, *Animal,* 111–14.

Weintraub, Linda. "Carolee Schneemann: Interspecies Eros." Seppälä, Vanhala, and Weintraub, *Animal,* 129–33.

Williams, David. "Book Review." *Performance Research* 5.2 (2000): 136–38.

6 Watching Eyes, Seeing Dreams, Knowing Lives

Marcus Bullock

What do we see when we look at an animal? What do we see especially when we look at an animal that has long had its place in the human world, such as a horse, a dog, or a cat? These animals are a part of our history, and one part of any answer to such a question will come down quite directly to a historical interpretation. Yet it is also true that they themselves are strangers to our history, just as they are strangers to our language. The look that they reflect back to us reminds us that in them we encounter something alien to the historical moment, though it may be difficult to see past the layers of apparent familiarity. Animals may not participate in the world of human speech, but the muteness that shrouds their senses always accompanies us in the realm of our language. Whatever else we may establish in the realm of language about them, despite all our convictions, all our knowledge, all our reasoning, we have to acknowledge that we are looking at something that eludes our ability to form a concept. Therefore, unless we refuse to look at all, the muteness of an animal also imposes a moment of muteness on us.

When we look at a horse, we see a body capable of motion and coordination much like our own. We also see movement and muscular expression that reveal responses to senses like our own. Moreover, this silent echo of our own being also exists in another plane of knowledge beyond the capacity of sight. The great power of those bodies, those full muscular forms that so far excel our own in their mass and capacity, can also share their command of the world with us. We can know their mobility almost as they know it, tactilely and kinesthetically, when we ride and run with them, when the blow of their hooves on the ground is also our moment of touch with the ground beneath us. This endows the image of the horse with a special quality that makes us see it differently from other animals, and this we find reflected in artistic representations of the horse. Even without humanized and sentimentalized distortions, the animal itself embodies something as deeply longed for and admired as the most formally accomplished representation of the human body.

This, naturally, also influences how we represent the association between the human figure and other animals. Despite the ancient bonds between men and dogs, that particular pairing lends itself very easily to representations of estrangement. To take only one instance, which could stand for countless others in modern literature, Albert Camus draws on the clarity with which we see nobility in the image of a horse and rider, but he develops the more ambiguous domesticity between a man and his dog to demonstrate how this more intimately formed connection can thwart our powers of self-perception. The air of self-containment in the power of a horse preserves a degree of dignity in our association with it. Our overfamiliarity with a canine companion produces a density that distorts our emotions or an obscurity in the way we manifest this tie between human and animal lives. For Camus, this lack of reflective clarity introduces us to part of the framework by which he represents the darkness and confusion from which disease-bearing rats bring their chaos into the hidden basis of human society.

In his first novel, *The Stranger*, we find the figure of Salamano and the mangy dog with whom he lives in perpetual irritation and miscommunication. The American translator explains that he corrected the earlier British rendition of "Il était avec son chien"—"As usual, he had his dog with him"—to "He was with his dog" in order to make clear that this would have been the phrase used to say he was with his wife, and thereby underlines the grotesque misassociation that the subsequent narrative will illustrate (vi).

Camus's counterpart to this miserable reality in a human-animal relationship appears in the endless dream that preserves the integrity of Joseph Grand in *The Plague*. This most minor of petty officials contrives to preserve dignity, and even honor, in an almost impossibly marginal existence. He succeeds in part by his perpetual creation and re-creation of a single image: a beautiful woman on a sorrel horse.

Yet these commonplace representations only draw on the predetermined potential of such images. Closer and more critical reflections on them occur more rarely. D. H. Lawrence's short novel *St. Mawr* contains one of the most intriguing instances of this development in modern literature. In that story, the presence of an extraordinarily powerful and noble bay stallion comes to stand as a dangerous rebuke to modern ways of seeing. When he is first introduced to us in the story, we learn that the animal has previously killed two of his riders. He will subsequently cripple or disfigure two more young men who enrage him by their modern incapacity to "meet him half way" (29), as they are told they must. The man who can find this point of balance comes from another world. This is Lewis, the Welsh groom for whom the English-speaking inhabitants of Britain are recent interlopers, and whom the narration describes as "aboriginal" (29) in Britain by contrast with them. He speaks to the horse in Welsh to calm and reassure him. He scarcely speaks to his employers at all, and when he does, this only deepens the sense of distance between him and the exchanges of a modern discourse.

Mrs. Witt, whose daughter Lou first finds and buys St. Mawr, notices the peculiar effect produced by the close bodily connection between Lewis and the stallion. She comments to Lou, "Isn't it curious, the way he rides? He seems to sink himself into the horse. When I speak to him, I'm not sure if I'm speaking to a man or to a horse" (35). Any appearance of directness in Lewis's speech to them always gives way to the impression of his watching them from a distance, which they compare to the gaze of a "human cat" (68), elusive, indifferent, elsewhere, watching with phosphorescent eyes "from the darkness of some bush" (27). Both the horse and the man who shares in his separateness work a similar fascination. Those who have fallen out of the blind faith in modern times, a faith exemplified by Flora Manby and her crowd, experience this distant presence as a look that reaches them from another world. Despite the impediment of their respective situations, Mrs. Witt even proposes marriage to Lewis as a means to cross the barrier of that look. But there is no way across.

Walter Benjamin defines that special quality of any serious work of art, which he calls its "aura," as our "investing it with the capacity to return our gaze" (188). Speaking quite literally, we can of course say that animals have the capacity to return our gaze as well, although to draw a parallel between animals and works of art on that literal basis would achieve nothing more than an empty play on words. Yet precisely because the gaze of an animal refuses us so much compared to the look we exchange with another person, to whom we may speak in expectation of a reply, we can say that the animal glances back not only literally but figuratively, through a capacity with which we must invest it, as well. The figure in this case may be illustrated by a line from Rainer Maria Rilke's poem "Archaic Torso of Apollo."

Rilke discovers the change wrought on an ancient Greek sculpture when he contemplates it in its ruined form. The head and limbs have broken away and vanished. Only the half turn in the torso survives as the expressive gesture of the fragment that remains. With the eyes lost, the stone has been disfigured as a representation of a humanized god, but it gains a different figural power in compensation. Now that we cannot look into a carved imitation of a face, we cannot imagine an idea in the represented mind. The figure no longer addresses us in the concept that we might construct in our minds as the "meaning" or the "statement" of the work. Without the eyes, Rilke discovers, "there is no place that does not see you" (165, translation modified). The result defeats interpretation, and perhaps even history.

This stone form bears only a tenuous relationship to the statue carved by its original sculptor to express the vision of his times. Reworked by the damaging accidents of other times, it stands at the very brink and outermost limit of human ideas. The consequence, Rilke's poem concludes, leaves us in a unique position. Because we cannot assimilate this form to a concept or an intention, it leaves us mute. Because we cannot take refuge in the defensive assimilation and distancing of an interpretation, it catches us unprepared and penetrates to another level of response. The poem concludes, "Du mußt dein Leben ändern"— you must change your life.

In something like the same sense, there is no part of an animal that does not look back at us. There is no part that does not remind us that there is something, a life, an existence that in some way echoes our own, but which remains always behind what meets our gaze, elusive, impossible, unimaginable. Our life undergoes a change from that experience of seeing an animal too. This may amount only to a small and fleeting color in our mood, but in that transient state, we recognize that we are only a part of ourselves. The part that we name and form as a concept cannot hold on to what we share with the elusive animal condition, wholly absorbed into its body and its senses.

And just as the resituated experience of awareness in Rilke's poem leads to a changed life, so it does in Lawrence's novel. Specifically, the two women, Mrs. Witt and her daughter Lou, find themselves alienated from the men of their contemporary world. Though each responds somewhat differently, and neither finds an altogether convincing new way of living, none of the former distinctions and hierarchies holds up under the weight of what they have discovered in Lewis and St. Mawr. When Lou and her mother argue about the qualities of mind that are so vividly absent in Lewis but that seem necessary to lift a man from the merely commonplace, Lou remarks,

> It seems to me there's something else besides mind and cleverness, or niceness or cleanness. Perhaps it is the animal. Just think of St. Mawr! I've thought so much about him. We call him an animal, but we never know what it means. He seems a far greater mystery to me than a clever man. He's a horse. Why can't one say in the same way, of a man: *He's a man*? (69)

When she asks her mother again how she can be so impressed by the ability of men to think, Mrs. Witt replies, "Perhaps I'm not—any more" (69).

In order to deepen our sense of the mystery, Lawrence falls back on the language of pagan mythology. Much discussion ensues invoking the idea of a pre-Hellenic Great Pan, and an esoteric mode of seeing that requires one to open up the third eye. While this is somewhat labored in the context, it does enable the narration to draw a contrast with the modern eye, whose grand expertise lies in observing surfaces. Lou's husband, Rico, that "almost fashionable" artist, sees only what he would paint (27, 78). The dean of the local vicarage exclaims, "The modern Pantheist not only sees the God in everything, he takes photographs of it!" (78).

So let us come back to that initial question. What do we see when we really look at an animal? Certainly not just what we observe. To illustrate the difference between the faculties of seeing and observing, let me take the example of an animal appearance that entails a much greater sense of distance than the horse. In Ernest Hemingway's *The Old Man and the Sea*, we find a story built on an encounter with a creature from the deep ocean. Unlike the apparent familiarity of a horse, the spectacle of a great fish still carries the shiver of the uncanny that grips us in the clash between two domains of life that can scarcely even engage one another. Hemingway explores the effects of strangeness when such a creature emerges into the realm where we as humans see things in our own terms. That strangeness makes us realize that those are not the only terms in which a world may be seen. When Santiago, the old fisherman, has caught a first glimpse of the huge marlin and has the image of it in his mind, he attempts to picture the fish's world as an image made by alien senses. "I wonder how much he sees at that depth, the old man thought. His eye is huge and a horse, with much less eye, can see in the dark. Once I could see quite well in the dark. Not in the absolute dark. But almost as a cat sees" (67).

His attempt to bring the creature closer by imagining it through the supposed greater familiarity of the horse and the cat does not last long. The strangeness that he comes back to again and again reasserts itself when he has captured the fish and brought it alongside his boat. Then he acknowledges that "the fish's eye looked as detached as the mirrors in a periscope or as a saint in a procession" (96).

Yet this, too, closely parallels the way Lawrence's characters come to experience the presence of St. Mawr, as opposed to the way they come to know one another and the reader comes to know them. The human figures appear before us in each case through a description of the gaze with which they meet the world and one another. The horse's eye, by contrast, only communicates its ancient distance and its opacity. St. Mawr's look reveals only that it holds something we cannot know, as when Lou feels a "ban on her heart" from the horse's glance: "St. Mawr looking at her without really seeing her, yet gleaming a question at her" (24). The apparent explicitness of the mutual recognition of persons who look each other in the eyes eludes the exchange with these animals, but we sometimes seek and find an alternative immediacy in a different mode of contact. For Santiago, a fish caught with a hook and line, like the horse trained to accept a saddle, communicates its strength through the sense of touch. This renews an ancient and yet commonplace connection that any angler knows when a fish first fights against the line and sends that throb of an unseen life up to quiver in the patient fisherman's hands.

The Old Man and the Sea was written by a man whom many regard as a master of observation, yet Hemingway also brings this text before us in the role of a seer. Because he seems to flatter a contemporary world that still lived in the conviction that it represented itself through the powers of modern observation, we see his seeing as embodying all that makes us suspicious about the seer from the point of view of the observer.

When the fish first breaks the surface, Hemingway's observation looks equal to all the importance of this meeting between two species, for it certainly registers the full beauty of the animal's form.

> The line rose slowly and steadily and then the surface of the ocean bulged ahead of the boat and the fish came out. He came out unendingly and water poured from his sides. He was bright in the sun and his head and back were dark purple and in the sun the stripes on his sides showed wide and a light lavender. His sword was as long as a baseball bat and tapered like a rapier and he rose his full length from the water and then re-entered it smoothly, like a diver and the old man saw the great scythe-blade of his tail go under and the line commenced to race out. (62–63)

Hemingway presents a vision in the ostensible form of an observation. We can identify that as an essential part of his literary aesthetics, and, of course, the aesthetics of a whole domain of literary practice beyond him. But if we observe

his observation closely, we see how the vision shows through in the careful di-
lation of observed time. The event itself, the breaking into view, begins at the
close of the first sentence, and would have been over, the fish would have already
vanished again, in the time it took to read the second sentence. The sentence
that follows recapitulates the spectacle in a different rhythm. It is read to us
from the memory that immediately succeeds the dazzled moment left behind
by a flash of life. Hemingway relies, as always, on repeating his characteristic
"and," which seems to convey the neutrality that merely links its observations
without intercession, and here occurs three times. By that device he conjures up
the simplicity of an eye that embraces each facet of its prey in innocent wonder,
and he identifies himself through this language as a man pure enough in his
spirit to give them all their full and equal due. The fourth and last sentence of
that paragraph has expanded the spectacle to include the vanishing moment. It
swells to exactly five times the length of the second, taking in five "ands," and
offering five similes with objects from the array of human tools, weapons, and
sports—the sword, the baseball bat, the rapier, the diver, and the scythe.

By the time the fish has disappeared beneath the water again, he has already
lingered long enough to have been buried under a mass of manly language.

Once the fish has plunged back into the deep, the text moves us from the
narrator's observations to the effect that this appearance has on a man engaged
in the great struggle that such a spectacle announces. Santiago's powers of sight
have already set him apart as strange among fisherman. Now we learn what he
has seen:

> He is a great fish and I must convince him, he thought. I must never let him learn
> his strength nor what he could do if he made his run. If I were him I would put in
> everything now and go until something broke. But, thank God, they are not as in-
> telligent as we who kill them; although they are more noble and more able. (63)

If the fish were capable of exact observation he would know how to snap the
line that joins him to his captor, but the senses that open up his own realm,
where he hunts and battles in the darkness of the deep, tell him nothing of this
way to freedom. The advantage on the side of the fisherman lies in the great
order of things that permits the play of strength to give way to the play of
minds. The story invites us to contemplate this as the reenactment of a timeless
rite. Santiago travels out alone, far from the sight of land, and not even the boy
Manolin has permission to sail with him and watch his battle. Without panoply,
without the crowd, without any witness or outward show at all, the labor of this
fisherman at his daily calling also resembles a corrida. What the fisherman sees
in his work recalls the confrontation in the bullring between the power of the
bull and the grace of the matador. The mute nobility of the beast accrues to the
man when both look together on the face of their death.

Santiago observes the spectacle of life in the fish and perceives it as an object
of beauty. When he looks beyond the first impact of the spectacle, he discovers
the place he and the creature share in the order of the universe, and declares he
sees it as an object of love. "Fish," he says, "I love you and respect you very much,

but I will kill you dead before this day ends" (54). What this signifies to him does not set him at odds with the idea of death. On the contrary, it sets him at peace in the understanding of his own transformation, the struggle for which might, each time, entail his own death instead of the fish's. That is why he sees the fish as his brother, the two of them linked in this fate just as much as they are separated by their different places in God's work.

Though the theological ideas here in Hemingway's formulation, just like the arcane mythology in Lawrence's *St. Mawr,* provide somewhat clumsy devices, they do reveal the separate vision through which an animal makes itself present to the eye inspired by a fuller concern. Santiago knows he is bringing his fish to market, but he is aware of the fish's dignity in himself as well as his value. Lawrence draws out the same thing by contrasting Lou's awe at the horse's nobility with the way her visitor, the Honorable Laura Ridley, responds to "St. Mawr's breeding, his show qualities." Lou has no interest in this critical measure and comparison. "Herself, all she cared about was the horse himself, his real nature" (174). Lou does notice that Laura has a kind of "reverence" for the show qualities, but their discussion shows that Laura is stirred by an aestheticism rooted in fashion, just like Rico's. Everything depends for them on display and possession. Therefore, like Rico, she belongs to the party of those who would happily see the horse shot. Their consciousness of beauty does not represent an awareness of the horse's real nature, but only of the narrow way in which his distinctiveness might reflect on their own vanity in the eye of society. This narrower perspective on themselves also prompts the longing in their particular modernity to preserve their absolute physical safety.

The sophistication of Lawrence's insight into modern sensibility keeps us from easily separating "observing" and "seeing" as stable opposites that might invite us to simply prefer one over the other. We could have mistakenly resorted to the distinction between some imaginary notion of scientific observation, that reports only something called "data," and an affectively richer sphere of aesthetics. This would be primitive in itself, and it would not help us understand how we experience the presence of animals. It is, of course, true that we can represent animals as mere objects of observation and data collection. It is also true that we could contrast this form of representation with the way animal images appear in works of art, or as the rhetorical resource for our most sacred texts. In either case, however, the world of animals themselves and our encounter with it have already fallen into a characteristic form of distance. I want to retrieve them from that distance.

If we restrict ourselves to the function of observer, approaching animals as assemblages of data, we quickly end up asking questions that would be entirely alien to a human mind in any other role. Where is the datum that animals feel pain? Where is the fact that distinguishes them from elaborate self-replicating machines? Why does their existence or extinction matter if it has no impact on what we want? Is there any other form of wanting than ours in the world? Any other right? Any other wrong? From a certain philosophical point of view, these questions may even be asked about human children before they acquire the power of speech and the ability to form concepts that accompanies this power. It is possible to argue that without those forms of consciousness that operate through concept and speech, we cannot rationally confirm for other beings the meaning that we attach to the term "pain" through our speech and recollected experience. Though we may observe the behavior of writhing and violent efforts to escape the source of hurt, and though we may hear screams, we cannot claim to observe the suffering that this might mean for a being like ourselves.

On the other hand, taking another faculty of the senses as our measure, we can say we see the suffering. To deny the spectacle of pain in an animal looks to such a capacity of sight like a horrifying blindness. If we look at an animal's suffering without the shield of this blindness, we register the pain not just as present but as a torment to our own senses. Though it may not rest on a concept, the vision strikes at us in an echo of suffering within our own bodily substance. The philosophical question "How do we know that animals feel pain?" rephrases itself under the force of that experience. It no longer means "Without an answer to that question, we are entitled to doubt that they do." Instead, it opens a much wider question about ourselves: "By what mysterious route does this intense experience and certainty of vision enter into us, so that the wish to deny it strikes us as a monstrosity?"

If by "seeing" we mean "knowing" all that the faculty of sight tells us about what actually meets the eye, then we have to go to war with our faculties to keep from seeing more than mere observation permits. There is, of course, a clear history of such a war, and the conflict has raged for a long time. Although I do not want to attempt a just account of the origins and the course of this war, I do think it may be about to draw to a close. We may be about to find a new assurance in the face of what our senses show us of the world as the picture becomes more complete through a greater sophistication in our technologies of observation. If that end does come about, then the entire character of the conflict may come into view without the need for elaborate interpretation. The war as I mean it here really consists in nothing but this contradiction buried in the basis of our senses, and the elaborate worlds of illusions that we, as an anomalous species, create and permit to entice us into adopting them as our natural habitat.

To limit the question to the simple matter of pain closes out the chance to consider more accurately just what it is that we see when we look at other species. To entertain the possibility that they do not feel pain moves the question too rigidly toward the issue of what we feel as the measure of what they do not. Instead, I would like to turn the comparison to the larger framework of those worlds that come into being through our senses and those of other animals. Those are what the animals of each kind—those with speech, and those without it—know as their natural habitat. Looking at the issue from this perspective shows that while we may be only an inessential part of their habitat, they are an indispensable element of ours.

That might seem an odd proposition in the light of the enormous destructive impact that the human presence has on animal worlds, but as long as we remain out of sight and sound and scent, they do not think about us. If we vanished, they would not miss us. We are simply a disturbance in their lives that would have no existence if it ceased to occur. On the other hand, we are such a disturbance in our own lives that only the rarest self-mastery can ever bring a moment's peace. The world that an animal takes in through its senses has formed in the course of an evolution that selects relentlessly against any waste, any diseconomy, any inefficiency that confronts an alternative. The senses therefore attune themselves only to those elements in an animal's world to which response has a net benefit. Like us, an animal does not live in "the world" as it "is," but only in the midst of those phenomena that its particular forms of awareness gather up and collect together for their combined relevance to its existence. Unlike us, they arrive at this attunement between themselves and their perceptual sphere through the mediation of their evolved bodily efficiency. The senses that measure out the space an animal inhabits, and the resources and dangers that space contains, serve the limbs and muscles that transform this space into a world without superfluity. Each aspect of this world—awareness, space, and mobility—exists only as an exact correspondence to the others. An animal existence consists solely in its ability to move through that world with the precision that keeps pace perfectly with all that the creature needs to approach or avoid. Neither the fear or repugnance that moves it away from a threat, nor the desire or hunger that moves it toward a necessity, can exceed the value this element bears for the animal's life as a representative of its form.

Scientists' observations can suggest remarkable structural correspondences between animals' responses and our own. Our reactions to a threat can be traced to the functioning of an organ in the brain called the amygdala, which evidently governs the same responses in other vertebrates. Similarly, the pleasure we take in certain activities results from the secretion of the neurotransmitter dopamine, which occurs as well in animal brains under precisely comparable circumstances. And yet the very study that permits us to learn of this similarity also establishes it as a profound form of difference between them and us. When we investigate the way they inhabit their environment in such precise harmonies and lasting forms, we respond to these phenomena as a world of beauty. Yet this world contains the knowledge of lives as well as the beauty of things. Those lives manifest an action of the senses that is simpler than, yet nonetheless akin to the knowledge we have of them. The insight into a living existence elicits a concern for its preservation connected with concern for our own existence, and valued beyond either beauty or a use in the human world to which it might be sacrificed. We not only see the elegance of harmonious outward shapes, but also arrive at an inkling of life as a perceptual system itself, so that we "see" through the sensitivities of response, selection, and adaptation that connect all organisms. That is to say, we derive a deep value from our understanding that the world of an animal is not the world that we construct for ourselves, with its constant excess and confusion. Our sensory apparatus burdens us, as reflexively conscious beings fully aware of time and change, with the inescapable demand that we bring that excess to order by all the conflicting means of language. We have to evolve and communicate explanations that may take a huge variety of social and expressive forms, from mythic symbols to material techniques, from theological to economic systems, from the conventions of a scientific to those of an artistic representation.

Where the images of animals that Lawrence and Hemingway create before us take on a religious cast, we can discern both a certain kind of anthropomorphism and a resistance to it. The animals are brought closer to the human world by their representation as spiritual beings, but their nobility and their distinction also keep them at a definite distance. In this discreet balancing of the two domains, we can experience an essence of all religious construction in our views of the cosmos. A religious image both anthropomorphizes the forces and activities of the material universe, by connecting them directly to the forces and acts of human choice, and questions any representation of humankind that restricts itself to a demonstrable, material anthropology. A spiritualized universe that we identify as the home of our inner longing to know ourselves calls on us to be more than the limited beings that we see in immediate reflection on the observed material of life. This sense of a call in those phenomena of an animal's beauty and presence by which it echoes our longings prompts the idea that we are indeed more than what we observe of ourselves.

In the realm of ordinary speech, animal life reflects a view of our own nature in two very different lines of metaphors. Some express a higher character symbolized by the beauty and power of these other beings; some express the lower aspect of compulsions and crude appetites. We can see ourselves as eagles or vultures, lions or jackals, bees or sloths. Yet such expressions limit us to quite ordinary levels of perception. Through the values embodied in these metaphors, we can judge ourselves either to have appropriately aspired to the ideal images of a social order or to have lived oblivious to the responsibilities of an actor on the full stage of human dignity before the tribunal of social judgment. In each case, however, we can also detect the thinness of observation that has gone into these conventional ideas. It does not require a great imaginative effort to discover ourselves through the knowledge we might acquire about animals.

The religious or spiritual perspective opens up the close framework of that conventional judgment, and the literature that explores this more complex realm of experience changes the language through which writing can make an animal presence accessible to the reader. The first order of concern in such literary exploration is that the text should place the rule of commonplace social motivations secondary. The threat to human life that both Lawrence and Hemingway portray in the encounter with animal powers is a clear point of separation between the more lowly interests pursued in ordinary social existence and a rediscovery of a deeper personal experience. The indifference developed by the primary figures in each story to social acceptance and economic advantage acquires its fullest significance in their equanimity before the risk to human life. The reflection on ourselves in the relationship to animals then invites us to see ourselves against the background of a more distant but also profoundly alluring alternative form of human life. Whether this human transformation is actually possible remains a religious question. We are entitled to be skeptical, since the religious traditions most illuminated by the symbolic presence of animals have fallen in stature. Nonetheless, what we contemplate in *St. Mawr* and in *The Old Man and the Sea* does not suffer at all by that diminution of a religious tradition. On the contrary, what falls away in tradition only returns us all the more strongly to the aesthetic and the experiential spheres.

Such questions then become more philosophical, and they require answers in the philosophy of literary representation. Though Lawrence and Hemingway may differ ideologically in critical ways, they come quite close together on this issue of literary purpose. The intense encounter with the beauty of an animal moves the protagonist in each case out of the orbit ruled by a barren social order, and into a finer, fuller consciousness; clearly the texts offer this as a path the reader should want to follow. That most people will indeed want to emulate this turn in the direction of a living experience because they already recognize something of it in their own lives does not mean that it must be real or true. The artistic technique of modifying an image to overcome its oft-repeated, and perhaps narrowly conventional, associations may revivify it and make it more effective, but that in itself does not indicate all we could mean by reality and truth.

We need to reflect once more on the idea that the spiritual significance of these narratives depends on sets of imagery in which the world of appearances both moves closer to the familiar domain, to evoke something we recognize in human experience, and at the same time moves these appearances further away, so we understand the opening, expanding realm of the human that must acknowledge something that is not itself in order to find more of itself. That is, such narratives must both anthropomorphize and resist the narrow meaning of anthropomorphism. In order to engage their readers, they have to address them in an established idiom of language and character. But if they are too indulgent in this direction, they run the risk of bringing their images too close to what we already know and can interpret confidently, as when we foreclose meanings within the observable correspondence of a simple metaphor. In order to reach out further they have to change the language and alter what we expect of the potential in human nature. If they go too far, they may be neither understood nor believed.

Since these alternatives require plausible resolution, we have to concern ourselves with the possibility that this literary representation of animals really does only conjure up its own particular obliviousness, another kind of forgetting and self-deception, or even a willful self-intoxication. Philosophical questions compel us to distinguish between the real and the irreal, the true and the false, the trustworthy and the deluding thought. The answer to those questions will certainly lie in the activity of the storytelling itself. In that activity, it is not the images themselves that carry the primary meaning, but the movement through them in which we experience the intangible coherence of the story itself. The force of that narrating, which preserves the essential distance in animal images like St. Mawr or the great marlin and brings its motifs together to constitute them in their depth, preserves a certain insecurity. In the end, it seems distinctly reticent about its own ideology, as though at least partly unsure about what it has seen.

Certainly, we find important alternatives to the literary figuration of the animal as an appearance through whose deep contemplation we leave behind our crass human misconceptions. Franz Kafka pursued a very different exploration of ways in which animal and human experiences might be found to resemble one another, and we should concede that his was a more rigorously imaginative enterprise. His narratives invent a form whereby a reflection articulated in a non-human character conducts the reader through a condition of language that drains away all the hope embedded in either domain of metaphoric usage, whether we thereby take animals as emblems of a higher or a lower condition within our cosmos. Kafka's reappraisal of those established figures of speech and the imagination on which they rest constitutes a definitive achievement in his work. His re-creation of literary language has established a new framework within which we acknowledge that his questions must be posed before any modern reading. Even so, it may be even more important to us to show that we can subject writings like those of Lawrence and Hemingway to Kafka's full critical force, and still derive something vitally beneficial to us from them.

Kafka's work depends on an immovability in the human condition. His stories are fraught with a striving for goals that always recede and revelations or transformations that only reveal the nothingness by which one state distinguishes itself from another. Behind that spare, ascetic prose which observes the movements of human understanding so closely, there stands a constant element determining the meaning of all the speech and behavior it represents, whether this turns upward toward a higher striving, or sinks down and resigns itself to the fall into more abject states of the human condition. The pessimism in Kafka's narrative order refuses to permit human choice to direct any change in that condition.

In this conception of life, freedom itself has no meaning and no value because the world reasserts its sameness everywhere, no matter what one might undertake in choosing a way through it. Even Karl Rossman, the naively hopeful protagonist who sets out with abundant youth and strength to change his life in *Amerika*, soon arrives at an experience that contradicts everything the grand tradition of the bourgeois age in writing taught us was a value and a human desire. "Yes, I'm free," Karl remarks, and then the narrator adds, "and nothing seemed to him more worthless than his freedom" (133).

In the short story "A Report to an Academy," Kafka speaks through the character of Red Peter, an ape who by heroic efforts of self-improvement has acquired the capacity for reasoned exposition. Red Peter offers only one definite conclusion from the unique vantage of this movement from life in the African forest to the company of learned humans. The power of speech and the new forms of knowledge have served only to secure "a way out" from confinement in a cage; he refuses to embrace the word "freedom" for his new condition (285). That word stands at the center of every reflection by humanity on itself in the tradition of Enlightenment that has embraced Red Peter and his achievements, but he cannot regard this concept as anything but the primary instance of blindness and self-deception in which all such efforts flow together and agree. That is to say, the community of these enlightened men has come together to share in a myth about themselves.

The blindness that fills them with confidence in the dignity of their knowledge and their community depends exactly on the workings of the mode of enlightened anthropomorphism itself. The ape who comes before them will either speak as they do, and thereby announce his freedom and his dignity as one who has risen by the labors of reason to the condition of humanity, or he will fall short of that elevated status, and confirm the difference between man and beast. By refusing to claim this state of emancipation and assimilation, while still making clear his capacity to grasp the concepts that would seem to confer this status as a mode of consciousness, he reveals a perspective in which the human species exemplified by the academy now appears to have enclosed itself in a kind of intellectual zoo.

Though it might seem paradoxical to argue this way, the literary innovations in Kafka's writing offer a more devastating hypothesis when considered philosophically rather than aesthetically. The reason for this lies in the Western philosophical tradition's rigid dependency on setting man apart as the rational animal. To infringe in any way on this essential distinction means putting the basis of human identity in question. Thus Theodor Adorno remarks in his "Notes on Kafka" that "Instead of human dignity, the supreme bourgeois concept, there emerges in him the salutary recollection of the similarity between man and animal" (270). The similarity that emerges here works to "salutary" effect because it disrupts the carefully managed modes of distance and association in which enlightened language controls the admissible and inadmissible correspondences between a conscious and a non-conscious creature. The metaphors drawing animals into the language by which we know ourselves limit the connections to specific, observable characteristics that operate for us in the realm of freedom. As long as our qualities of character depend on the operation of freedom, then objective models by which we can exemplify those qualities serve our freedom too. Adorno will not leave his comment at any such reassuring point, however.

Human dignity in its bourgeois form, resting on its Enlightenment basis, has in Adorno's view fallen utterly into self-deception because it has elevated capacities that contribute to rational domination above those that determine human happiness. In consequence, our civilization has based its relationship to animals on the difference between creatures with and those without this power of abstract knowledge and the ability to dominate nature it confers. All our thoughts that follow in this pattern carry us further away from understanding any other constitution of identity, and have alienated us especially from the quality of a relationship with another based on a shared capacity to feel pain, privation, or joy and fullness of spontaneous vitality. The notion of anthropomorphism thus divides into two very different kinds of illusion according to two different images of human nature.

On the one hand, we find it entirely commonplace to take the term "anthropomorphism" to name that familiar inclination to project human feelings into animal experiences. We do not have great difficulty in recognizing the temptation for what it is, when we find ourselves looking for a sensation or emotion we know in ourselves in order to interpret a posture or gesture in an animal. Anything that strikes us as "expressive" in the behavior of another creature makes us pick something in our human vocabulary of appearances to which we can see a correspondence, and then let that "expression" speak to us as though we had made a reliable translation from one bodily form to another. On the other hand, Adorno's critique of Enlightenment rationalism can also find an equivalent error in the opposite mode of interpretation, in the rigidly assured vision that sees nothing but the operation of human knowledge anywhere in the universe. That steadfast refusal to see expressiveness anywhere merely becomes another species of anthropomorphism, should we turn so intently against the other temptation as to insist on hearing only silence and seeing only empty matter in the language of animal forms.

To deny any possibility that animals do experience something, even though we may not have the resources in our linguistic imagination to represent it, projects human knowledge beyond its true domain. This projection will always qualify as myth, even if it presents itself in the guise of rejecting myth in its more overt, animist, manifestation, as happens in the materialist or empiricist view that denies animals a place in the domain of real experience. What especially complicates the issue here derives from the aesthetic quality of myth, or the exceedingly delicate relationship between knowledge and beauty. The assurance of a full knowledge always brings with it the subjective sense of something in the world having yielded itself up to us and arrives with an escort of pleasure. A form that lets us see a manifest order in appearances, therefore, can, by its subjective allure, take on the role or place of knowledge. Conversely, if we reflect on that pleasure, and on the desire that prompts us to embrace it as though it revealed a "truth" in the objective world, we also discover ourselves already present in those graceful or reassuring appearances. We reveal the desire for that sense of knowledge as the agent of a preformed harmony projected on the world around us. This means that aesthetic experience takes two forms, each quite different in its relationship to myth, though we find ourselves somewhat misleadingly limited to the single term "aesthetics."

One perspective within aesthetics leans more toward accentuating the fact that we may voluntarily suspend disbelief in order to receive pleasure from an artistic representation. This invites us to reflect on the artifice involved in composing a harmonious form that compensates us for the roughness of the world. We may remain quite content to find nothing more in its sensual qualities than an escapist irreality that complements, for the span of its passing moment, our living in that real, material sphere. The other perspective takes art more "seriously" as the representation of an ideal that appearances in the material world can only approximate. This idealizing view of form gives art substantial authority as a source of truth. Philosophical analysis of aesthetics must deal with the tangled interconnections between these aspects in order to free us of myth and illusion, and the example of the Frankfurt school reveals how illusory any single position taken on these two outlooks may turn out to be in the course of a history that constantly changes the quality of our experience and the focal points of authority in human knowledge.

"Enlightenment has always taken the basic principle of myth to be anthropomorphism, the projection onto nature of the subjective" (6), Adorno and Horkheimer state in the opening section of *Dialectic of Enlightenment,* their gloomy critique of human history written at the end of World War II. In so doing, they begin to rework this principle according to their finding that, in high modernity, enlightenment has dialectically absorbed its own opposite and become myth itself. By identifying human essence with the abstract subject, with the refined consciousness that emerges entirely in control of itself, and therefore as "free," this striving against myth has anthropomorphized human nature. The paradoxical anthropomorphism of man has simultaneously alienated the substance of human happiness in bodily existence from this essentialized subject, and then re-created an entire cosmology on the assumption that this subject is the measure of all things, whatever aspect of knowledge it confronts: "Oedipus' answer to the Sphinx's riddle: 'It is man!' is the Enlightenment stereotype repeatedly offered as information, irrespective of whether it is faced with a piece of objective intelligence, a bare schematization, fear of evil powers, or hope of redemption" (7).

This second version of anthropomorphism includes animals too, which it has likewise alienated from the experience of life in the overwhelming Enlightenment urge to achieve the "disenchantment of the world" by the "extirpation of animism" (5). That ideology has effectively removed the figure of the animal, or any relationship to animals, as a place of philosophical contemplation by providing this total view of the world as the Other of enlightened knowledge. The stifling weight of rationalist oblivion falls over the world of animals and our capacity to reflect on them just as it numbs us to our own potential life. Adorno compares the remaining concern for animals in a modern sensibility to the amusement medieval society derived from human grotesques. "Animals are only remembered when the few remaining specimens, the counterparts of the medieval jester, perish in excruciating pain," he claims, and holds out little hope that wild creatures will long survive in this hostile human environment (251). He predicts, "They will be completely eradicated," since "[t]he earth, now rational, no longer feels the need for an aesthetic reflection" (251).

If the prediction looks as though it might be quite mistaken in view of the concern for animal preservation that has suddenly come to the fore in our contemporary culture, we can also see a mistaken view of aesthetics at work here. The literary judgment that would certainly place Kafka above Lawrence or Hemingway takes the integrity of the work of art in itself, the complete predominance of the medium over its material, as the highest priority of artistic composition. And when we marvel at the beauties of Kafka's work, we acknowledge the perfection of a particular extreme in aestheticism. The beauty of the work depends in no way on the beauty of what it portrays. Kafka brings before us the unalloyed bad news of the ugliness of our world, the pervasiveness of filth, the ubiquity of cruel impulses, the disappearance of love. His animals bring us that news too. Their lives exemplify an alternative version of that very same condition of our world. The fluid power of Kafka's representation in prose derives from the ease of portraying one kind of hopelessness through another. Our inability to rise above the condition of animals permits them to enter into our forms of experience without their encountering any contradiction, just as we can imagine waking up like Gregor Samsa in *The Metamorphosis,* occupying another body but discovering how little we actually had to lose in our erstwhile humanity.

Adorno admires the ruthless artistic will behind this merciless exposure of all that we find missing in ourselves. What he may have underestimated in himself appears in the philosophical extremism of self-denial entailed in such a willful hopelessness. This goes hand in hand with the reluctance to accept worldly beauty as a source of pleasure or hope, and the insistence everywhere in Adorno's writing that only in the ascetic rigor of an autonomous work of art can we see the last living refuge of human emancipation. He mistrusts any pleasurable recollection of nature as a regression mediated by the system of consumable phantasms. This mistrust is as "salutary" as what Kafka shows us in our similarity to animals, because it unveils a comparable illusion. It places us at the center of a system of meanings by which we have been induced to surrender a real, vital place in the world where we live. This constitutes the pattern he notes in the progress of universal assimilation to a repressive culture. "The history of civilization is the history of the introversion of sacrifice," he and Max Horkheimer declare in *Dialectic of Enlightenment.* "In other words: the history of renunciation" (55).

Yet to renounce the aspect of art that directs our aesthetic reflection to beauty in the world certainly looks like another version of such sacrifice. And though we perhaps should take the unattractive aspect of Lawrence's and Hemingway's social or political views as connected quite closely to what they saw in the realm of nature, the vision of beauty and nobility itself that they bring before the reader does not necessarily lead us to those views at all. What they do achieve by foregrounding the vital experience of an encounter with another kind of being leads us to feel what it might mean to renounce the authority of the reigning social order altogether.

In the context of modernity, this returns us to animals as a very powerful source for the meaning of freedom, just as we can imagine that paleolithic cave paintings represented a liberating source of power to their creators. In both cases we understand that this may well not change the world, but it does change us who live in it, and this understanding has a real benefit in giving us pause before we continue the great enterprise of modernism that threatens to destroy the world in the process of changing it.

For this reason, it does behoove us to carefully contemplate our experiences with watching animals around us as we attempt to formulate a philosophical understanding of ourselves. The thesis of a human essence constituted in the rationality of a transparently self-present subject figures in philosophy as the counterpart of an animal essence constituted in the functions of an organic machine. The former necessitates the latter. Armed with the appropriate convictions, we can see both exemplified around us everywhere. Freed of that prejudice, we can see a quite different world. In particular, what we see of our world when we look at it closely and well always requires of us that we change our life. While we may see ourselves as the rational animal, we also see ourselves as the irrational animal. We can surprise and disappoint ourselves.

We live by time and memory, plan and expectation. The capacity to constitute and reconstitute the idea of ourselves in response to this always incomplete vision forms the basis of the human capacity to tell stories, which is certainly a unique and distinguishing quality of the human situation in the world. Of course, this too brings us before the difference between what we see and what we observe. A story makes us see a consistent unfolding pattern of reality through a set of circumstances, events, or actions distributed so that we may observe them in time. The element that reveals itself through these events has a remarkable quality indeed, in that it may be "true" even though the story itself is fictitious. The story constitutes something we know to be real in our world, and real about ourselves, our character and relations with others in the world, because of the way the events mirror forms and patterns in the world that do not reveal themselves to observation alone, but to the creative eye of the seer. The worlds that we do see in this visionary process are "true" insofar as we do indeed live in them, and do cohere in a productive way with what appears to our sensory intuitions. Yet, as Wordsworth notes in his lines written on the Wye above Tintern Abbey, these proceed from an active faculty in our senses working in concert with the purely receptive awareness, as "what [the eye and ear] half create / And what perceive" (99).

The question of how we place an animal within a story therefore goes to the heart of the question of what we learn by watching these other beings live their lives and by contemplating what they reveal to us as knowledge about their living, which cannot know the call to change that we hear all the time, as we did in the closure of Rilke's poem. *The Old Man and the Sea* and *St. Mawr* both respect the limits of a true storytelling in that, while the animals remain absolutely central to the story because their presence motivates all the change and development in the human characters, the narration nonetheless keeps them peripheral by showing us nothing of their interior changes. So it is with the experience of other animals, just as it is with works of art, or even with an inanimate object of natural beauty. These participate in our language in their own way, neither by the transmission of information to us nor as the consequence of an intention, but by calling on a recognition in us who have the gift of language and meanings.

And what we must see in them, unless we refuse to make a due account of what we observe, comes down to the desire to live, and to live by desires that inspire their strength and their energy according to pleasures and pains, fears and appetites, tastes and disgusts. This recognition precisely constitutes the difference between what language means to us and what it would mean to any kind of machine that we are able to conceive of at present. Animals may not choose as we do, and cannot change their lives as we can, but they do choose in their own way. We can watch an animal in a situation where its desire for an attractive morsel of food lures it on, while the risk of exposure to a possible predator holds it back, and perhaps its awareness of us taking in its quandary and a nearby rival also act together to inhibit a decision. These desires and awarenesses manifest themselves in the signs of its attention to each, small motions to direct its senses, listening, looking, sniffing the air, the readiness of its body quivering with contrary eagernesses. Then, out of all this inner response and bodily reaction, it reaches a decision. It does not compute statistical probabilities and plot a course, but seizes on a line of action in a way that we understand. It embraces the moment, and its whole system commits itself, just as we commit ourselves when we arrive at a decision, centering our identity around it and feeling it as the fullness of our moment. The difference in our choice remains, because we do have the capacity to recollect and replay the process within larger and larger fields of meaning, but the basis of this moment, filled with the experience of interacting desires, remains a characteristic of life we share with the animal.

No matter how sophisticated we make a machine that deciphers what is said to it, or what it observes in a visual field, nothing in this interaction rises to equal the content of an experience. Though we might very easily program into it responses that mimic any kind of human experience, such as dislike of pain, fear of death, moral outrage, disgust, or pursuit of a pleasure, these responses can only originate in the person who constructs the machine. Interpreting behavior as gesture or speech as expression can never qualify as anything more than an anthropomorphization of the device. Nothing that we find there proceeds from pleasure or pain, fear or desire, but only from a sequence of controls that rests on the internal neutrality of the device itself, and the external disposition of a mechanical design deliberated on and made elsewhere, out of sight.

Then what is it we recognize in an animal? Not the language in which we might try to couch what we see. There we necessarily fall into parallels, metaphors, displacements. The life and the energies that an animal reveals to us act on our language by the difference between its existence and ours, separated from us by desires, pleasures, impulses, pains, fears, or repugnances that it alone knows directly, and our own. We recognize that the creature is not indifferent to itself, whereas a machine does not know any difference within itself. No matter how complicated a chess-playing IBM machine might be, it does not want to win. It simply executes a winning algorithm. Indeed, one reason why it can win is that it exploits predictable features of a human player who does want to win and does fear defeat. A creature without language cannot look at the world in terms of a metaphysics, but its impulses organize its world around it into an ordered field of impressions, resemblances, and warnings, which is what language does when animated by our own desires. A machine equipped with an adequate input of data concerning its surroundings still does not form the center of any world of which it might construct an "image." The information figures as a world only in relation to the design and execution of the machine's purposes, whose center lies outside, in the plan of its designer.

The machine registers a mock neutrality in the gathering of information, though it remains subordinate to a purpose that lies outside itself. Its objective power can so overwhelmingly fascinate the persons and subjective forces that pursue this purpose that this instrumental focus in human activity re-creates every concept of human life in its image. The authors of *Dialectic of Enlightenment* look at the mechanization rampant in their world by analyzing a philosophy they regard as the direct ideological reflection of a material and social history. Adorno and Horkheimer connect the forces of industrial capitalism with the Enlightenment notion of objective knowledge, and declare, "This kind of neutrality is more metaphysical than metaphysics" (23). In the age of information processing, their formula would define a further stage. The machines of our day do not simply mimic work in the material domain by shaping and assembling objects. They supplement our ability to remember, to compare, to represent, and to predict. Yet without the distinction between pain and desire, between pleasure and fear, their mimicry does not center on an essential difference between existence and non-existence. Indeed, without these qualities of the body in pleasure and pain, it develops no center at all. The neutrality of this machine existence reveals to us what it is that we continue to need from the place once occupied by metaphysics.

The dominion of facticity under the philosophical framework of Enlightenment resisted the religious dogmas that had previously ordered the universe around a divine center of knowledge, and displaced this point of authority with the knowledge possessed by the abstracted consciousness of a theoretical human subject. These were both alienating forms because neither placed what we observed around us in a just relationship with what we cannot keep from seeing beyond that narrow horizon. The decentered realm of electronically managed information undoes this unjust formulation of our identity within a set array of master narratives grounded in metaphysics. Without the ability to focus a motivated structure of desires, however, it cannot supply our need for a truer story. It would seem that the only story that can be told through the appearance of a mechanical contrivance in human life will simply relate the inability of this encounter to produce the changes we require of ourselves to make our life present and visible to us. Our similarity to animals does not make us more human. The task of realizing our humanity remains entirely our own. Our similarity to animals simply protects us from sinking into our resemblance to machines.

Without the example provided by animals before our eyes, we as a species might be unable to imagine a state beyond the constantly re-created series of delusions in which our existence consists. It has often been pointed out that for us fear is not objective but endemic, because we know the world as a set of concepts that always includes threats and dangers. It is no less true that we are prey to endemic desires that depend on the same fantastic relationship to our world. The world is just as capable of seeming to blossom with promises made by unseen potentials. Our world, in short, is a habitat without proportion or measure.

All the estimates we make within its visible horizons are thrown into confusion by our existing far beyond our senses in realms of language and transcendent computation, in dreams and savage excesses of speculation.

This endemic disturbance runs right through the distinction I made by separating observation and seeing. These are the two halves of the world that we inhabit, as animals of a unique quality. The dream of retreating into the realm of other animals, the dream that haunts the two women who own St. Mawr, may promise relief from the bitterness of life frozen among mere objects. Yet this estranges us no less from the kind of animals that we are. It fetishizes one part of our being as a species, and thereby destroys our powers in the other half and leaves us just as homeless as before. But in this moment of our present history, we have begun to formulate our relationship to animals differently than at all other times. This may have come about only because we can imagine a world without animals, now that our powers of destruction have grown so monstrously. Another reason, however, may be that we are learning to observe our own observations ever more clearly. That has permitted us to see them as ever less the expression of a transcendent rationality and ever more the expression of a quality rooted in our nature just as deeply as that of any other animal. The phenomena of our world lie prefigured in our body and our senses too. To speak from this understanding allows us to describe ourselves as "the animal that speaks" in a new sense, and if we remember Lou's question and dare to say "I am an animal" as we would say "I am a man" or "I am a woman," that too now has a real meaning.

Bibliography

Adorno, Theodor W. "Notes on Kafka." *Prisms*. Trans. Samuel Weber and Shierry Weber. Cambridge, Mass.: MIT, 1967. 243–71.
———, and Max Horkheimer. *Dialectic of Enlightenment*. Trans. John Cumming. New York: Continuum, 1999.
Benjamin, Walter. "On Some Motifs in Baudelaire." *Illuminations*. Ed. Hannah Arendt. Trans. Harry Zohn. New York: Schocken, 1969. 155–200.
Camus, Albert. *The Plague*. Trans. Stuart Gilbert. New York: Knopf, 1948.
———. *The Stranger*. Trans. Matthew Ward. New York: Random, 1988.
Hemingway, Ernest. *The Old Man and the Sea*. New York: Simon and Schuster, 1995.
Kafka, Franz. *Amerika*. Trans. Willa Muir and Edwin Muir. New York: New Directions, 1962.
———. "A Report to an Academy." *The Metamorphosis, In the Penal Colony, and Other Stories*. Trans. Joachim Neugroschel. New York: Simon and Schuster, 1995. 281–93.
Lawrence, D. H. *St. Mawr*. New York: Knopf, 1925.
Rilke, Rainer Maria. *New Poems*. Trans. J. B. Leishman. New York: New Directions, 1964.
Wordsworth, William. *The Prelude: Selected Poems and Sonnets*. Ed. Carlos Baker. New York: Holt, 1954.

7 ... From Wild Technology to Electric Animal

Akira Mizuta Lippit

In one of a series of interviews with David Sylvester, Francis Bacon makes the following assessment of photographs of animals in slaughterhouses:

> I've always been moved by pictures about slaughterhouses and meat. . . . There've been extraordinary photographs which have been done of animals just being taken up before they were slaughtered; and the smell of death. We don't know, of course, but it appears by these photographs that they're so aware of what is going to happen to them, they do everything to attempt to escape. (23)

The precise nature of Bacon's sentiment ("moved," he says) remains ambiguous in this account—a delicate mixture of arousal, pity, indignation, fascination, and even love, perhaps. (Kojève has suggested that human beings love animals for the same reasons, and with the same capacity, that they love the dead.[1]) Regardless of the exact emotion, Bacon appears overwhelmed by a torrent of sensation. He imagines not only the intentions of the animals ("attempt[ing] to escape") but also their odor—the smell of animals—intermingling with the more ambient "smell of death." In his reaction, Bacon appears to shift from the vantage point of a witness to that of a participant. He has slipped into the diegesis of slaughter, somewhere between slaughterer and slaughtered—in a kind of "photographic ecstasy," to invoke Roland Barthes's phrase (119). Bacon appears to identify with the soon-to-be-slaughtered animals.

In the same exchange with Sylvester, Bacon continues his identification: "We are meat, we are potential carcasses. If I go into a butcher shop I always think it's surprising that I wasn't there instead of the animal" (46). Bacon's rediscovery, in front of the photograph, of a totemic and sacrificial economy is here revealing. The factuality of meat appears to be an issue here not for the ecstatic fact that it reveals—that corporeality continues beyond the threshold of mortality; that even after "we" die, we remain as animal facts, artifacts, and after the fact—but, rather, for the excessive force of its facticity. The force of this fact, here in the form of a photographic image, seems to "puncture," to use Barthes's idiom, a register that was not stimulated by the initial phenomenon. In this case, the olfactory organs appear to have been activated in the wake of a visual perception. Bacon, as it were, smells the photograph, both its brutality and its facticity. In smelling the photograph, the animals in the photograph, Bacon assigns the two media a referent, a body. Between the figure of the animal and the materiality of the photograph, Bacon posits a mutual body of movement, intention, odor. He inscribes a locus of identification, a will, a dynamic that in turn moves him as spectator. In the figure of impending death, evoked by its smell in the photograph, Bacon sees himself. Between the slaughterhouse and the butcher shop, the animal and the photograph, Bacon finds himself slipping into the temporality of identification—Bacon sees himself in the place of the animal subject, an imminent corpse: "We are potential carcasses."

Bacon's utterance exposes a problem concerning animals and identification—namely, the assumption that human beings cannot identify with animals. Since the animal possesses no discernible subjectivity, the human subject cannot rediscover itself in the place of this other. While a human being can project anthropomorphic characteristics onto the animal or experience emotions (such as pathos or sympathy) in response to its being, an impenetrable screen—language—divides the loci of animal and human being. If Bacon has indeed effected an identification with this image, then where does one locate the source of Bacon's identification: in the animal or in the photograph? The question raised by Bacon's uncanny sentiment addresses the possibility that the combination of the animal subject and the photographic image alters in some essential fashion the structure of identification. Although both the animal and the photograph impede the dialectical flow of subjectivity (the effects of fascination and ecstasy result in the termination of the subject), as an assemblage, as a *rhizome,* animals and photographs appear to found and animate an entirely other topology: one that allows for an economy of the gaze, identification, and becoming.

Identifications result from encounters with sensual excess: a subject identifies with an image when that image exceeds the visual register and penetrates the polymorphic body. By entering the phenomenal *mise en scène* of the slaughterhouse, Bacon not only advances from the secondary to the primary level of identification, he crosses the frame that separates reality from photographic reproduction, nature from the artwork, animality from humanity, and life from death. Identification is a mode of becoming, of mimesis, a method by which the ego assumes the properties of an other.[2] Identifying with the animal is part of the process of becoming-animal. The photograph's carnal reach that Barthes has noted appears to have touched Bacon here. He is becoming-animal, becoming-photograph, and the two becomings are inseparable. As with Barthes's characterization, the photograph's *punctum,* its sensational excess, initiates the spectator's feelings of identification. The spectator feels, in front of certain photographs, a sense of identification with them, although this experience indicates a hallucination. It is this perverse trespass that facilitates identification. The photograph effects the subtle shift of being into an entirely other world. Of the slight distances traversed in photographic replication—the otherworldly vestige of its remove—Bacon reflects, "I think it's the slight remove from fact, which returns me onto the fact more violently" (30). Photography brings the spectator violently back to the reality of the real that appears, in the first instance, elsewhere.

The remove from fact that restores an awareness of that fact's truth to the spectator is not a feature exclusive to photography. The slight remove from fact is also a property of the animal look. Even at a distance, the animal look, as John Berger notes, elicits a type of recognition from the gazing subject.[3] Krafft-Ebing offers this glimpse into the psychopathology of "zooerasty," or pathological bestiality: "In numerous cases, sadistically perverse men, afraid of criminal acts with human beings, or *who care only for the sight of the suffering of a sensitive being, make use of the sight of dying animals, or torture animals, to stimulate or excite their lust*" (125, emphasis added). In Krafft-Ebing's account, animals supplant the immediacy of human encounters. Animals, or rather images of animals, mediate the violent act. The image of the suffering animal facilitates, in this case, a move beyond the conventions of so-called human behavior. One can exceed the permissible limits of human violence by violating the image of the animal. The animal look does not terminate the momentum of identification, but rather deflects it into another economy. Thus displaced, identification mutates in kind: it no longer adheres to or circuits through the subject, but opens onto another space of identification unimpeded by the responsibilities of reason, language, and consciousness. By projecting the vector of identification into the animal world, the subject avoids what Derrida refers to as "the call that originates responsibility" ("Eating Well" 112). This impossible identification with the animal can be likened to an ingestion of the animal, invoking the transferential logic of sacrifice.

By consuming the animal in identification, the subject undergoes a becoming-animal in an effort to disappear from the realm of responsibility. As evidenced in anthropological research, the act of eating is commonly conceived of as a method of incorporating and becoming the other. Funeral feasts as well as most other forms of ritual consumption are directly linked to the notion of becoming-other, of harnessing the powers of the other (especially in the case of animals) by consuming its flesh. And because the animal inhabits an apolitical world, identification with this sphere exempts humanity from participating in human ethics. The sacrificial eating that concludes the act of identification—an identification that transpires in the act of eating—transfers onto humanity what Adorno calls the dying animal's "manic gaze" (105). The human being becomes other, and thus returns violently to the fact of itself, by consuming the animal, by engaging in what Derrida terms "eating well" (*bien manger*) ("Eating Well" 96). Regarding animal consumption, Derrida questions the constitutions of humanity and humanism in relation to nature and the animal:

> The subject does not want just to master and possess nature actively, in our cultures, he accepts sacrifice and eats flesh. I would ask you: in our countries, who would stand any chance of becoming a *chef d'État* (a head of State), and of thereby "acceding to the head," by publicly and therefore exemplarily declaring him- or herself to be a vegetarian? . . . The *chef* must be an eater of flesh (with a view, moreover, to being "symbolically" eaten himself). . . . In answering these questions, you will have not only a scheme of the dominant, of the common denomina-

tor of the dominant, which is still today in the order of the political, the State, right, or morality, you will have the dominant schema of subjectivity itself. ("Eating Well" 114)

One now recognizes in Bacon's ecstatic shudder over the sight of slaughtered animals—a vision that seems to jolt him violently back into the place of the real—the source of his excitement. A convergence between eating and seeing, the two modes of phenomenal consumption that involve animals and photographs, brings these two entities together in a sacrificial economy. At the same time, this economy institutes a specific temporality. In an x-ray view of the animal, Bacon sees the animal's essence, its future. Animals can only appear as matter—meat—because they possess no discernible identity. The photographs thus show the spectator the future of the animal. Such disclosures are, for Barthes, a salient feature of photography. In *Camera Lucida,* Barthes also stumbles upon this realization, this anterior future tense of the photograph:

> In 1865, young Lewis Payne tried to assassinate Secretary of State W. H. Seward. Alexander Gardner photographed him in his cell, where he was waiting to be hanged. The photograph is handsome, as is the boy: that is the *studium.* But the *punctum* is: *he is going to die.* I read at the same time: *This will be* and *this has been;* I observe with horror an anterior future of which death is the stake. By giving me the absolute past of the pose (aorist), the photograph tells me death in the future. What *pricks* me is the discovery of this equivalence. In front of the photograph . . . I shudder, like Winnicott's psychotic patient, *over a catastrophe which has already occurred.* Whether or not the subject is already dead, every photograph is this catastrophe. (96)

Like the condemned man in Barthes's photograph, animals are both dead and alive, suspended in the photograph.

Regarding Bacon's professed identification with animal carcasses, one must consider the metaphorics of eating, the act or gesture that circumscribes most sacrificial ceremonies. Concerning the relationship between eating and all phenomenal activity, Derrida asserts, "For everything that happens at the edges of the orifices (of orality, but also of the ear, the eye—and all the 'senses' in general) the metonymy of 'eating well' (*bien manger*) would always be the rule" ("Eating Well" 114). Looking at the photograph, Bacon recognizes his own death in the future by internalizing the corpse as a narcissistic fact. All photographs, as Barthes says, are of future corpses.

What is shared by animals and photographs is a crypt, in Nicolas Abraham and Maria Torok's sense, in which the antitheses—animal and technology—are united without, however, producing a sublation. A secret synthesis that cannot signify, the photograph brings into focus the alliance between animal and technology. One realizes that animals and photographs often produce the same phantasmatic and liminal effect, disrupting the flow of figurative language. Animals are, in this sense, fleshly photographs.

Animals can be seen as predecessors of photography, the two joined by their particular look. Animals expose what remains an unnameable aspect of fascination. And if Bacon transfigures the dynamic of the animal into that of its flesh, then he is only accelerating what is already embodied in the image of the animal, its corpse. For André Bazin, the idea of embalming distinguishes the *technē* of the photographic image.[4] But how do animals facilitate this transit between corporeality and photography? Why are animals the ideal subjects of photographs? One remembers that animals have never properly belonged to any ontology. Derrida explains that in traditional philosophy, "there is no category of original existence for the animal" ("Eating Well" 111). This is because animals have been excluded from the essential categories that constitute being. Nonetheless, animals have sustained the existence of every category of being as essential supplements: without belonging to any ontological category, animals have made those categories possible by situating their borders. It is only in the imaginary topology of the photograph that one is able, perhaps, to perceive or discern the animal world. It is in such an exposure that the animal enters, for the first time, the phenomenal world. The animal look can be seen as a continuation of the photographic look. One recognizes that similitude in Bacon's photograph as a kind of *punctum*.

Technology

> When, as a child, I was told about the invention of the telegraph the question which most interested me was, wherever did the swallows gather for their autumn migration before there were telegraph wires?
>
> —Karl von Frisch[5]

Cinema, which builds hallucinatory space, can be seen as a cryptic topography in which animals and the reproductive media become one another, forming a Deleuzian rhizome. As animals began to disappear from the phenomenal world, they became increasingly the subjects of nineteenth- and twentieth-century reproductive media. In 1872, Eadweard Muybridge began the task of bringing movement to the still image, of arousing an animate vitality from the catatonic corpus of photography. In his work, animals began to move across the reproductive media. In sequence after sequence, the photographer's animals pushed against the cryogenic frames of a fixated medium until they seemed to surpass the limits and enter the interstices, creating the semblance of motion—persistence of vision.

Muybridge's collection, *Animals in Motion,* which was begun in 1872 and appeared in copyrighted book form in 1899, and its sequel, a zoomorphic treatment of human bodies, *The Human Figure in Motion* (copyrighted in 1901), display the fascination with which animals and animal movement captured the photographic imagination.[6] Muybridge, whose technical contributions also include the zoopraxiscope, a projected version of the zootrope, has similarly been positioned along a track that leads to the discovery of the cinema. What is remarkable in Muybridge's work, what immediately seizes the viewer's attention, is the relentless and obsessive manner in which the themes of animal and motion are brought into contact—as if the figure of the animal were predestined to serve as a symbol of movement itself. The movement of Muybridge's animals, across first the frames and then eventually the screens of a new industrial landscape, not only aided the advent of a new mode of representation—cinema—it also introduced a new way to transport information from one locale to another; from one forum to another; one body to another; one consciousness to another. By capturing and recording the animals' every gesture, pose, muscular disturbance, and anatomical shift with such urgency, Muybridge seemed to be racing against the imminent disappearance of animals from the new urban environment. Distinct from the stillness of photography, cinema added the possibility of electric animation.

Christian Metz explains the relationship of cinema to time. In film, "everything is recorded (as a memory trace which is immediately so, without having been something else before)" (43). Film is a parasite of the real, of the now, which is why it cannot be restricted to a theory of language. Films are graphic constitutions of the real in the past: they record and then exhibit the past—a past that has never been, in the first instance, present. In this sense, film projects a totally imaginary relation to time and the world. For Metz, "[w]hat is characteristic of the cinema is not the imaginary that it may happen to represent, but the imaginary that it *is* from the start, the imaginary that constitutes it as a signifier" (44). As such an a priori expression of the imaginary, film calls into question the primacy of language in the constitution of the human world. This is because film simulates, in a manner to be discussed shortly, the effects of transference—the rapid movement of affect from one entity to another.

This is not to suggest that film determines a facile dialectic with language, that the properties of image, sense, and projection can be readily placed against those of language, reason, and presence. Film does not replace language, for it cannot exist without it. Film displaces language, exposes the abyss that threatens to engulf every semantic signification. Film parasitizes language, much as the animal does, drawing into its imaginary panorama that which remains undisclosed in discursive transactions. Cinema is a parasite.

The *Oxford English Dictionary* places the first known use of the word *anthropomorphism* with the meaning "attribution of human traits to animals" in the second half of the nineteenth century. (Until this referential shift, the word was used to indicate mistaken attributions of human qualities to deities.) It was during the nineteenth century, with the rise of modernism in literature and art, that animals came to occupy the thoughts of a culture in transition. As they disappeared, animals became increasingly the subjects of a nostalgic curiosity. When horse-drawn carriages gave way to steam engines, plaster horses were mounted upon tramcar fronts in an effort to simulate continuity with the older, animal-powered vehicles. Once considered a metonymy of nature, animals came to be seen as emblems of the new, industrial environment. Animals appeared to merge with the new technological bodies that were replacing them. The idioms and histories of numerous technological innovations, from the steam engine to quantum mechanics, bear the traces of an incorporated animality. James Watt and later Henry Ford,[7] Thomas Edison,[8] Alexander Graham Bell,[9] Walt Disney, and Erwin Schrödinger,[10] among other key figures who contributed to the industrial and aesthetic shifts of the late nineteenth and early twentieth centuries, found uses for animal spirits in developing their respective machines, creating in the process a series of fantastic hybrids. Cinema, communication, transportation, and electricity drew from the actual and phantasmatic resources of dead animals. Technology, and more precisely the technological instruments and media of that time, began to serve as virtual shelters for displaced animals. In this manner, technology and ultimately the cinema came to determine a vast mausoleum for animal being.

Crypt

A paradox surrounds animal death. Since animals are denied access to the faculties of language, they remain incapable of reflection, which is bound by finitude and carries with it an awareness of death. Undying, animals simply expire, transpire, shift their *animus* to other animal bodies. Of the complete absence of death among animals, Bataille asserts, "Not only do animals not have this consciousness [of death], they can't even recognize the difference between the fellow creature that is dead and the one that is alive" (2 and 3: 216). Animals thus function as the incarnation of a technological fantasy—perpetual motion machines. Thomas Sebeok writes of the animal, "Whatever else an animal may be, it is clear that each is a living system, or subsystem, a complex array of atoms organized and maintained according to certain principles, the most important among these being negative entropy" (159). From the viewpoint of both quantum mechanics and metaphysics, animals appear capable of moving at will between the states of life and afterlife. By facilitating such perpetual engines, animals brought technology to life—the animal spirits that entered into the technological body turned technology into a species. Machines might fail, suffer, experience the breakdowns of exhaustion and confusion, but animated machinery as a *technogeny* would survive the demise of individual apparatuses.

In *The Wolf Man's Magic Word*, Abraham and Torok argue that Freud was unable to cure his famous Wolf Man patient, Sergei Pankeiev, because the latter had formed a crypt, a radically other locus of subjectivity that could not be opened by the instrument of language.[11] Because the encrypted subject was not bound by language in the ordinary sense of the term (Pankeiev, according to the authors, had created a secret and absolutely singular idiom), he was not restricted by its finitudes. The Wolf Man was not mortal: he was incapable, according to the authors, of dying.

In presenting their theses on cryptic subjectivity—one that incorporates the other in its entirety without processing or integrating its alterity—Abraham and Torok further enable the logic of an "illicit union" (Deleuze and Guattari's term) between technology and the animal. In this context, modern technology can be seen as a massive mourning apparatus, summoned to incorporate a disappearing animal presence that could not be properly mourned because, following the paradox to its logical conclusion, animals could not die. It was necessary to find a place to which animal being could be transferred, maintained in its distance from the world. Abraham and Torok's maintenance system, in which the encrypted other is "crossed *out*," writes Derrida, "kept alive so as to be left for dead," resembles Heidegger's crossed-through animal world.[12] As Leibniz writes of animal death, "There is therefore no *metempsychosis*, but there is *metamorphosis*" (209). If the animal cannot die but is nonetheless vanishing, then it must be transferred to another locus, another continuum in which death plays no role. Animals must be transformed into cryptological artifacts.

The technological crypt resembles its psychic counterpart to the extent that both preserve the radically absent other in a state that can be defined as neither life nor death. Put another way, the structure of the crypt preserves the presence of an absent other that has never been present. Like Metz's cinema, which records everything as a memory trace before it has been anything else, the crypt presents the absent other before it has ever been present. This cryptological other has no world of its own, appearing for the first time as an absent other in the crypt. This is close to the complicated structure of world assigned to the animal. It is thus perhaps more than mere coincidence that Abraham and Torok find in the case of the Wolf Man the elements of a crypt.

Another factor links the crypt with technology and cinema. The crypt itself is an effect of psychic *technē*. The crypt, Derrida writes, is an "artificial unconscious"—the "Self's artifact" ("*Fors*" xix). Like Jean-Louis Baudry's notion of film, which induces an artificial psychosis, Derrida's crypt establishes an artificial unconscious—an artificial subjectivity.[13] Derrida describes the unnatural topology of the crypt: "Carved out of nature, sometimes making use of probability or facts, these grounds are not natural. A crypt is never natural through and through. . . . The crypt is thus not a natural place [*lieu*], but the striking history of an artifice, an *architecture*, an artifact" ("*Fors*" xiv). Crypts are, according to Derrida, technological in nature. They effect a technology of the subject. The brutality of the cryptic artifact transforms the self into other, even as it transforms the other into self or subject. In the crypt, in the artificial unconscious, the absent other becomes subject. Technology becomes a subject when it gains an unconscious; that artificial unconscious is established by the incorporation of vanishing animals.

Like the Wolf Man's irreducibly singular "verbarium"—his world composed of cryptonyms—technology also produces a structure and a site of communication that avoid conventional language. In its most basic manifestation, electricity determines the currency of technological communication. In the case of Freud's Wolf Man, the crypt is cracked by a dream, the Wolf Man's dream of "six or seven" white wolves on a tree.[14] In their analysis of this dream, Abraham and Torok suggest that the Russian word for window, *okno*, hinges upon a similar word for "'eye,' that is *oko* or *otch*, the root for its inflected forms" (*Wolf Man's* 33). The polysemic slide that Abraham and Torok discover is compelling not only because it marks the transition from organ (eye) to apparatus (window), perhaps even to the cinematic apparatus (lens), but also because it displays the method by which cryptonyms enter into the bodies of other words. Abraham and Torok write of the secret word or words that determine the crypt,

> The key word, no doubt unutterable for some reason, and unknown for the moment, would have to be polysemic, expressing multiple meanings through a single phonetic structure. One of these would remain shrouded, but the other, or several other meanings now equivalent, would be stated through distinct phonetic structures, that is, through synonyms. To make our conversations about this easier, we would call them *cryptonyms* (words that hide) because of their allusion to a foreign and arcane meaning. (*Wolf Man's* 18)

Those synonyms or cryptonyms construct an unconscious dimension within the words that carry their meanings. Secret words travel parasitically from one host body to another in a movement that closely resembles transference. (In fact, as Abraham and Torok note, Freud and Otto Rank fought over the Wolf Man's dream. Freud trusted its authenticity; Rank claimed that the dream was a product of the patient's transference with Freud.)[15] The logic and history of transference carry the trace of animal magnetism into the psychoanalytic field. Transference is the means by which non-verbal energy circulates within the world.

Sandor Ferenczi defines transference as "[the] means by which long forgotten psychical experiences are (in the unconscious phantasy) brought into connection with the current reaction exaggerated by the affect of unconscious ideational complexes" (36). Laplanche and Pontalis add "a particular instance of displacement of affect from one idea to another" (457). In *The Interpretation of Dreams,* Freud establishes the "indestructibility" of unconscious wishes or affects as the groundwork for transference. The reserved force of unexpressed desire is harnessed to various persons or acts that enter into the unconscious field and is released, in this manner, into the world. And because, like animals, unconscious wishes are indestructible, undying, they are recycled constantly throughout the world. Freud writes, "I consider that these unconscious wishes are always on the alert, ready at any time to find their way to expression when an opportunity arises for allying themselves with an impulse from the conscious and for transferring their own great intensity on to the latter's lesser one" (5:553). In this way, then, the affect is transferred from an inarticulate vehicle to a phenomenal word, act, or body. These affects operate like parasites, transferred by contact even when two conscious directives fail to reach an accord. Transference may thus take place even where communication does not—a kind of telepathy, or unmediated contact between metalinguistic forces.

In a footnote to the above passage, Freud follows the current of the perpetual unconscious affect through its metamorphic cycle. Of the perennial wishes that fill the reservoir of transferential activity, Freud writes,

> They share this character of indestructibility with all other mental acts which are truly unconscious, i.e. which belong to the system *Ucs.* only. These are paths which have been laid down once and for all, which never fall into disuse and which, whenever an unconscious excitation re-cathects them, are always ready to conduct the

excitatory process to discharge. *If I may use a simile, they are only capable of annihilation in the same sense as the ghosts in the underworld of the Odyssey—ghosts which awoke to new life as soon as they tasted blood.* (5:553n)

Unconscious wishes, like phantom animals, cannot be destroyed. They lie dormant until another source awakens them. Because animals are unable to achieve the finitude of death, they are also destined to remain "live," like electrical wires, along the transferential tracks. Unable to die, they move constantly from one body to another, one system to another.

Cinema

In this glum desert, suddenly a specific photograph reaches me; it animates me, and I animate it. So that is how I must name the attraction which makes it exist: an *animation*.

—Roland Barthes

"Cinema begins," writes Sergei Eisenstein, "where the collision between different cinematic measures of movement and vibration begins" (1:192). Eisenstein's assertion, like the writings of Dziga Vertov, Germaine Dulac, Antonin Artaud, and others who sought to discuss the new medium in ritual forms, bears the trace of an organic metaphor, an attempt to describe technological animation in animist terms. Many of Eisenstein's extensive writings on cinema are figured by an organic idiom. Describing the physiological status of cinema, Eisenstein writes,

This montage is not constructed on the *individual dominant* but takes the sum of *stimuli* of all the stimulants as the dominant.

That distinctive montage *complex within the shot* that arises from the collisions and combinations of the individual stimulants inherent within it,

of stimulants that vary according to their "external nature" but are bound together in an iron unity through their reflex physiological essence.

Physiological, in so far as the "psychic" in perception is merely the physiological process of a *higher nervous activity.*

In this way the physiological sum total of the resonance of the shot *as a whole,* as a complex unity of all its component stimulants, is taken to be the general sign of the shot.

This is the particular "feeling" of the shot that the shot as a whole produces . . .

The *sum totals* thus achieved can be put together in any conflicting combination, thereby opening up quite new possibilities for montage resolution.

As we have seen, because of the actual genetics of these methods, they must be accompanied by an extraordinary *physiological* quality. (1:183)

According to Eisenstein, the edit or montage of the scene is sustained only to a degree by the dominant stimulus of the shot. With each edit, a multiplicity of extraneous information is also carried across the discursive threshold. A series of minor or imperceptible edits accompanies every major edit. Vertov refers to these minute edits as intervals: "*Intervals* (the transitions from one movement to another) are the material, the elements of the art of movement" (8). The intervals are like unconscious thoughts or genetic codes, passing like secrets through the dominant semiotics of the shot. Vertov and Eisenstein argue for a biology of the cinema, whether seen as psychology or physiology: for an understanding of cinema as organism.

Referring to montage as hieroglyphic "copulation," Eisenstein extends the notion of film editing as a system that resembles the confluence of genes in organic reproduction. As each montage sequence within a film text combines with further and more detailed units of montage, the culminating effect of "their interrelationships . . . move[s] towards more refined variants of montage that flow organically from one another" (Vertov 191). The term *organic* in Eisenstein's usage moves beyond the idiomatic nuance of a seamless flow to denote, rather, the transmission of complex data from one shot to another—not all of which may cross the viewer's perceptual threshold. According to Eisenstein's fantasy, filmic shots, like genetic structures, comprise dominant and recessive traits: when they are crossed, certain features are exposed upon the surface of the filmic body while others perform a subliminal function, sustaining the linkage between shots. Cinema cannot exhibit all of its features: as with any genetic code, certain materials are made manifest while other information remains recessed. For Eisenstein, the filmwork gradually evolves into a body distinguished by an exterior and interior dimension, if not by a conscious and unconscious capacity. Eisenstein articulates the cinema virtually as a biological organism. The characteristics with which he describes the medium, "reflex physiological essence" and "higher nervous activity," apply to the filmwork itself rather than to its spectator. One senses in Eisenstein's cinema a biomorphic hallucination. Films exist here as complex organisms—they have become animal, or *animetaphor:* animal and metaphor, a metaphor made flesh, a living metaphor that is by definition not a metaphor, antimetaphor—animetaphor.

Eisenstein's animetaphor here functions as a technology, as do all instances of animetaphoricity. Despite the concept of nature it references, the animetaphor is itself profoundly unnatural, prosthetic, pressing the limits of world against the void. In an entirely different sense from that put forth by Descartes, the animal as figure functions as a technological trope, a technological index. The animal projects from a place that is not a place, a world that is not a world. A supplemental world that is, like the unconscious, like memory, magnetic in the technological sense. Thus when Freud insists that he does not know "what animals dream of," the statement is more than a rhetorical figure or illustration.[16] His inability to know this articulates concretely the very mechanism of dreams. One experiences the dream as a world under erasure; dreams mark the becoming-animal of human beings. And because dreams originate in a place that is not a place and arrive only as an originary translation or technological reproduction, they appear only within the frames of a mediating apparatus. Of Freud's dream logic, Derrida notes, "Everything begins with reproduction" ("Freud" 211). Derrida's mention of reproduction here refers not to the duplication of something that already exists, however, but rather to the introduction of something else, something other, through a technology of representation. Regarding Freud's struggle to inscribe a metaphor for the apparatus of memory, Derrida stresses the inevitable turn in Freud's rhetoric from biological (1895) to technological (1925) figures. The *Wunderblock,* a fantastic machine that is not without its animal traces, evolves in Derrida's narrative from the organic memory cell: "Metaphor as a rhetorical or didactic device is possible here only through the solid metaphor, the 'unnatural,' historical production of a *supplementary* machine, *added to* the psychical organization in order to supplement its finitude" ("Freud" 228). In a similar manner, the animal as metaphor, the animetaphor, supplements the dream, language, and world systems, providing an external source of energy that charges the machine. The animetaphor is in this sense both alien and indispensable, an electromagnetic spirit that haunts the unconscious.

Thought of as technology, as a *technē* that opens worlds, the animetaphor operates like a fabulous machine. Trace and memory, Nietzschean amnesia or Heideggerian erasure, the discourse on the animal reveals at its origins a technological *atopia*—a world that, as Derrida claims, always begins on the occasion of its reproduction.[17] Another apparatus, cinema, which arrives with psychoanalysis in 1895, provides, perhaps, the proper metaphor for the impossible metaphor, the animetaphor. The function of *unheimlich* reproduction, the vicissitudes of affect, the dynamics of animation and projection, the semiotics of magnetism, and the fundamental properties of memory can be seen as the basis of cinema, but also of the animal. Cinema is like an animal; the likeness a form of encryption. From animal to animation, figure to force, poor ontology to pure energy, cinema may be the technological metaphor that configures mimetically, magnetically, the other world of the animal.

One final speculation: the cinema developed, indeed embodied, animal traits as a gesture of mourning for the disappearing wildlife. The figure for nature in language, animal, was transformed in cinema to the name for movement in technology, animation. And if animals were denied the capacity for language, animals as filmic organisms were themselves turned into languages, or at least into semiotic facilities. The medium provided an alternative to the natural environment that had been destroyed and a supplement to the discursive space that had never opened an ontology of the animal. In a radical departure from the framework of nature, the technological media commemorated and incorporated that which it had surpassed: the speechless semiotic of the animal look. Animal magnetism had moved from the hypnotist's eye to the camera eye, preserved in the emblematic lure of cinema. As a genre, animation—from Oskar Fischinger's spermatic ballets to Walt Disney's uncanny horde—encrypted the figure of the animal as its totem. Thomas Edison has left an animal electrocution on film, remarkable for the brutality of its fact and its *mise en scène* of the death of an animal. The single shot of an animated film-elephant collapsing from the surge of electrical current brings together the strange dynamic of life and death, representation and animal, semiotic and electricity. It is emblematic of the uncanniness of the medium. In the filmwork, one experiences the convergence of a traditional opposition between nature and artifice, *phusis* and *technē*, animal and technology. Cinema, then, can be seen as the simultaneous culmination and beginning of an evolutionary cycle: the narrative of the disappearance of animals and that of the rise of the technical media intersect in the cinema. The advent of cinema is thus haunted by the animal figure, driven, as it were, by the wildlife after death of the animal.

Notes

This essay was first presented as a lecture to the Center for Twentieth Century Studies at the University of Wisconsin–Milwaukee. The text was later published in an expanded and revised form in *Electric Animal: Toward a Rhetoric of Wildlife* (Lippit). The present version preserves the scope of the original lecture. I am grateful to Nigel Rothfels and the Center for their contributions to this text, and to the University of Minnesota Press for the right to reproduce this excerpt from *Electric Animal*.

1. One need not be able to "recognize" a loved object, Kojève explains, to continue loving it. One loves "given-Being (*Sein*)" and not "Action (*Tun*)" or "Product (*Werk*)." Thus, he asserts, following Goethe, "one loves a man not because of what he *does* but for what he *is*, that is why one can love a dead man . . . that is also why one can love an animal, without being able to 'recognize' the animal" (244*n*).

2. Citing Lalande, Laplanche and Pontalis note that identification determines an "[a]ct whereby an individual becomes identical with another or two beings become identical with each other" (205).

3. Berger writes, "The eyes of an animal when they consider man are attentive and wary. The same animal may well look at other species in the same way. He does not reserve a special look for man. But by no other species except man will the animal's look be recognised as familiar. Other animals are held by the look. Man becomes aware of himself returning the look" (4–5).

4. "If the plastic arts were put under psychoanalysis," Bazin claims, "the practice of embalming the dead might turn out to be a fundamental factor in their creation" (1:9).

5. From *Dancing Bees* (4). In thanking his translator, von Frisch comments, "Suppose German and English bees were living together in the same hive, and one of the Germans found a lot of nectar: its English companions would easily understand what it had to say about the distance and direction of the find. Human language is not so perfect" (ii).

6. Both *Animals in Motion* and *The Human Figure in Motion* are derived from Muybridge's complete portfolio of photographs, *Animal Locomotion*, which was published in 1887. The most thorough analyses of "chronophotography," from Muybridge to Etienne-Jules Marey, have been conducted by Michel Frizot. See, for example, *La chronophotographie*.

7. One horsepower was calculated at 33,000 foot-pounds per minute in England, while a French *cheval-vapeur* was defined by the ability of a horse to lift seventy-five kilograms one meter in one second. Lewinsohn explains, "After James Watt had built his steam engine, he wanted to find out how much work

his machine could accomplish. The most impressive way of measuring work done was by comparing the engine's output with that of a horse. Physiological experiments revealed that a horse could work constantly at the rate of 22,000 foot-pounds per minute. This figure was arbitrarily increased to 33,000 foot-pounds per minute and called a 'horsepower'" (273). At the instant of its conception, then, the engine was already imagined as an equine crypt. In his chapter titled "Horsepower," Lewinsohn illustrates the assorted devices that resisted the disengagement of the horse from the engine.

8.   Attempting to impede the consolidation of competing AC (alternating current) distributing systems and advance his own DC (direct current) generators, Thomas Edison and his then assistant Charles Batchelor regularly demonstrated the execution of animals with high-voltage alternating current. Josephson explains, "The big laboratory at West Orange was the principal source of 'scientific' evidence purportedly exposing all those who were making and selling a-c systems to the public. There, on any day in 1887, one might have found Edison and his assistants occupied in certain cruel and lugubrious experiments: the electrocution of stray cats and dogs by means of high-tension currents. In the presence of newspaper reporters and other invited guests, Edison and Batchelor would edge a little dog onto a sheet of tin to which were attached wires from an a-c generator supplying current at 1,000 volts. . . . The feline and canine pets of the West Orange neighborhood . . . were executed in such numbers that the local animal population stood in danger of being decimated" (347). Edison's particularly brutal massacre of West Orange's animal population suggests an intensity of purpose: Edison seemed willing to eliminate the entire population of local animals to prove that alternating current was lethal. A kind of phantasmatic exchange can perhaps be seen in Edison's repetition compulsion: animals for electricity, life for power. One might speculate that animals, in Edison's laboratories, were reducible to pure force, *animus*, electricity. Josephson describes an incident in which Edison's assistant Batchelor accidentally received the animal's voltage: "In one of those sadistic 'experiments,' Batchelor, while trying to hold a puppy in the 'chair,' by accident received a fearful shock himself and 'had the awful memory of body and soul being wrenched asunder . . . the sensations of an immense rough file thrust through the quivering fibers of the body.' Though badly shaken up, he recovered in a day or two, it was said, 'with no visible injury, *except in the memory of the victim*'" (347). The episode highlights the strange connection between the space of animals, electricity, and psychic multiplication.

9.   Avital Ronell has connected Bell's research on sheep to the advent of the telephone. Troubled by the relative weakness of the sheep's reproductive system, Bell attempted to increase the number of their nipples. Ronell writes, "For Alexander Graham Bell the sheep take up a significance of affective investment of the same intensity as the telephone. He must multiply nipples, keep the connection going; they need to be kept from perishing" (338).

10.   The animal achieves a kind of vital superiority in Austrian physicist Erwin Schrödinger's 1925 "black box" demonstration of quantum mechanics. A cat is placed in an experimental crypt along with a radioactive nucleus and a cyanide capsule with a trigger device that will be activated by the nucleus's decay.

"After one minute," writes commentator-scientist Paul Davies, "there is a fifty per cent chance that the nucleus has decayed. The device is switched off automatically at this stage. Is the cat dead or alive?" According to the "overlapping waves" that represent different possible states of both the cat and its poison, the animal rests in its quantum casket simultaneously dead and alive. Davies concludes his commentary by psychoanalyzing the cat: "[I]t seems that the cat goes into [a] curious state of schizophrenia . . . and its fate is only determined when the experimenter opens the box and peers in to check on the cat's health. However, as he can choose to delay this final step as long as he pleases, the cat must continue to endure its suspended animation, until either finally dispatched from its purgatory, or resurrected to a full life by the obliging but whimsical curiosity of the experimenter" (131). In Davies's description, one returns to the strange metaphysics of Schrödinger's dead and alive cat, which allows one to conceptualize alternative, parallel worlds. Technology emerged as the memorial and aesthetic of an extinct animality. Ironically, within the technological frame, animals became subjects. Animal substance became the very possibility of the new technological environment. Schrödinger writes in his "philosophy": "If we consider our earthly environment, it consists almost exclusively of the living or dead bodies of plants and animals" (41).

11. Of the secret words that attest to the Wolf Man's subjectivity and crypt, Abraham and Torok assert, "What distinguishes a verbal exclusion of this kind from neurotic repression is precisely that it renders verbalization impossible" (*Wolf Man's* 21). See also Abraham and Torok, *L'écorce et le noyau*.

12. Derrida, *"Fors,"* xix. Of the animal world and its relation to language, Heidegger offers this prescription: "When we say that the lizard is lying on the rock, we ought to cross out the word 'rock' in order to indicate that whatever the lizard is lying on is certainly given *in some way* for the lizard, and yet is not known to the lizard *as* rock" (198). For Heidegger, the animal world and word are always under erasure. See, in this connection, Derrida, *Of Spirit.*

13. Moving from Bertram Lewin's hypotheses on the dream screen, Jean-Louis Baudry argues that the cinematic screen doubles as the maternal breast and forms, for the viewer, a type of surrogate maternity that simultaneously triggers regression and a kind of "artificial psychosis." Baudry writes, "Cinema, like dream, would seem to correspond to a temporary form of regression, but whereas dream, according to Freud, is merely a 'normal hallucinatory psychosis,' cinema offers an artificial psychosis without offering the dreamer the possibility of exercising any kind of immediate control" ("Apparatus" 315). The image of a nurturing film brings the apparatus closer to the threshold of animality, at least to mammalian animals. In the same anthology, see also Baudry, "Ideological Effects."

14. *"I dreamt that it was night and that I was lying in my bed. (My bed stood with its foot towards the window; in front of the window there was a row of old walnut trees. I know it was winter when I had the dream, and night-time.) Suddenly the window opened of its own accord, and I was terrified to see that some white wolves were sitting on the big walnut tree in front of the window. There were six or seven of them. The wolves were quite white, and looked more like foxes or sheep-dogs, for they had big tails like foxes and they had their ears*

pricked like dogs when they are attending to something. In great terror, evidently of being eaten up by the wolves, I screamed and woke up" (Freud, "From the History" 17:29).

15. See *Wolf Man's* 33, and the section "The False 'False Witness' and the Rank Affair," *Wolf Man's* 52–54.

16. "I do not myself know what animals dream of" [Wovon die Tiere träumen, weiß ich nicht] (Freud, *Interpretation* 4:131).

17. In "On the Uses and Disadvantages of History for Life," Nietzsche links the absence of language in animals not to deprivation but rather to a chronic forgetting, which results in the animal's envied appearance of happiness. Of the apparent happiness of animals, Nietzsche writes, "A human being may well ask an animal: 'Why do you not speak to me of your happiness but only stand and gaze at me?' The animal would like to answer and say: 'The reason is I always forget what I was going to say'—but then he forgot this answer too, and stayed silent: so that the human being was left wondering" (60–61).

Bibliography

Abraham, Nicholas, and Maria Torok. *L'écorce et le noyau*. Paris: Flammarion, 1987.

——. *The Wolf Man's Magic Word: A Cryptonymy*. Trans. Nicholas Rand. Minneapolis: University of Minnesota Press, 1986.

Adorno, Theodor. *Minima Moralia: Reflections from Damaged Life*. Trans. E. F. N. Jephcott. London: Verso, 1974.

Bacon, Francis. *The Brutality of Fact: Interviews with Francis Bacon*. Ed. David Sylvester. Oxford: Alden, 1975.

Barthes, Roland. *Camera Lucida: Reflections on Photography*. Trans. Richard Howard. New York: Hill and Wang, 1981.

Bataille, Georges. *The Accursed Share*. Trans. Robert Hurley. New York: Zone, 1991.

Baudry, Jean-Louis. "The Apparatus: Metapsychological Approaches to the Impression of Reality in the Cinema." Trans. Jean Andrews and Bertrand Augst. *Narrative, Apparatus, Ideology: A Film Theory Reader*. Ed. Philip Rosen. New York: Columbia University Press, 1986. 299–318.

——. "Ideological Effects of the Basic Cinematographic Apparatus." Trans. Alan Williams. *Narrative, Apparatus, Ideology: A Film Theory Reader*. Ed. Philip Rosen. New York: Columbia University Press, 1986. 286–98.

Bazin, André. "The Ontology of the Photographic Image." *What Is Cinema?* Trans. Hugh Gray. Berkeley and Los Angeles: University of California Press, 1967. 1:9–16.

Berger, John. "Why Look at Animals?" *About Looking*. New York: Vintage, 1980. 3–30.

Davies, Paul. *Other Worlds: Space, Superspace, and the Quantum Universe*. New York: Simon and Schuster, 1980.

Derrida, Jacques. "'Eating Well,' or the Calculation of the Subject: An Interview with Jacques Derrida." *Who Comes after the Subject?* Ed. Eduardo Cadava, Peter Connor, and Jean-Luc Nancy. Trans. Peter Connor and Avital Ronell. New York: Routledge, 1991. 96–119.

——. "*Fors:* The Anglish Words of Nicolas Abraham and Maria Torok." Trans. Barbara Johnson. Abraham and Torok, *Wolf Man's,* xi–xlviii.

——. "Freud and the Scene of Writing." *Writing and Difference.* Trans. Alan Bass. Chicago: University of Chicago Press, 1978. 126–231.

——. *Of Spirit: Heidegger and the Question.* Trans. Geoffrey Bennington and Rachel Bowlby. Chicago: University of Chicago Press, 1989.

Eisenstein, Sergei. "The Fourth Dimension in Cinema." *S. M. Eisenstein: Selected Works: Writings, 1922–34.* Ed. and trans. Richard Taylor. Bloomington: Indiana University Press, 1988. 1:181–94.

Ferenczi, Sandor. "Introjection and Transference." *Sex in Psychoanalysis: Contributions to Psychoanalysis.* New York: Robert Brunner, 1950. 35–93.

Freud, Sigmund. "From the History of an Infantile Neurosis." 1918. *The Standard Edition of the Complete Psychological Works of Sigmund Freud.* Ed. and trans. James Strachey. 24 vols. London: Hogarth and the Institute of Psycho-Analysis, 1953–74. 17:7–122.

——. *Interpretation of Dreams.* 1900. *The Standard Edition of the Complete Psychological Works of Sigmund Freud.* Ed. and trans. James Strachey. 24 vols. London: Hogarth and the Institute of Psycho-Analysis, 1953–74. 4:xi–338, 5:339–627.

Frisch, Karl von. *The Dancing Bees: An Account of the Life and Senses of the Honey Bee.* Trans. Dora Ilse. New York: Harcourt, Brace, 1953.

Frizot, Michel. *La chronophotographie.* Beaune, France: Association des Amis de Marey et Ministère de la Culture, 1984.

Heidegger, Martin. *The Fundamental Concepts of Metaphysics: World, Finitude, Solitude.* Trans. William McNeill and Nicholas Walker. Bloomington: Indiana University Press, 1995.

Josephson, Matthew. *Edison: A Biography.* New York: John Wiley and Sons, 1959.

Kojève, Alexandre. *Introduction to the Reading of Hegel: Lectures on the* Phenomenology of Spirit. Ed. Allan Bloom. Trans. James H. Nichols, Jr. Ithaca: Cornell University Press, 1969.

Krafft-Ebing, Richard. *Psychopathia Sexualis.* Trans. F. J. Rebman. New York: Physicians and Surgeons, 1925.

Laplanche, Jean, and J.-B. Pontalis. *The Language of Psycho-Analysis.* Trans. Donald Nicholson-Smith. New York: Norton, 1973.

Leibniz, Gottfried Wilhelm. "Principles of Nature and Grace Based on Reason." *Philosophical Essays.* 1714. Ed. and trans. Roger Ariew and Daniel Garber. Indianapolis: Hackett, 1989. 206–12.

Lewinsohn, Richard. *Animals, Men, and Myths: An Informative and Entertaining History of Man and the Animals around Him.* New York: Harper, 1954.

Lippit, Akira Mizuta. *Electric Animal: Toward a Rhetoric of Wildlife.* Minneapolis: University of Minnesota Press, 2000.

Metz, Christian. *The Imaginary Signifier: Psychoanalysis and the Cinema.* Trans. Celia Britton, Annwyl Williams, Ben Brewster, and Alfred Guzetti. Bloomington: Indiana University Press, 1982.

Muybridge, Eadweard. *Animals in Motion.* 1899. Ed. Lewis S. Brown. New York: Dover, 1957.

——. *The Human Figure in Motion.* 1901. New York: Dover, 1955.

Nietzsche, Friedrich. "On the Uses and Disadvantages of History for Life." *Untimely*

    Meditations. Trans. R. J. Hollingdale. Cambridge: Cambridge University Press, 1983. 57–123.

Ronell, Avital. *The Telephone Book: Technology-Schizophrenia-Electric Speech.* Lincoln: University of Nebraska Press, 1989.

Schrödinger, Erwin. *My View of the World.* Trans. Cecily Hastings. Woodridge, Conn.: Ox Bow, 1983.

Sebeok, Thomas. "'Animal' in Biological and Semiotic Perspective." *American Signatures: Semiotic Inquiry and Method.* Ed. Iris Smith. Norman: University of Oklahoma Press, 1991. 159–73.

Vertov, Dziga. "We: Variant of a Manifesto." *Kino-Eye: The Writings of Dziga Vertov.* Ed. Annette Michelson. Trans. Kevin O'Brien. Berkeley and Los Angeles: University of California Press, 1984. 5–10.

PART 3

Cultures of Animals

8 Unspeakability, Inedibility, and the Structures of Pursuit in the English Foxhunt

Garry Marvin

The foxhunt in England is a dramatic enactment of a set of relationships among foxes, hounds, horses, and humans. It is an event which both depends on representations of animals and has actively constructed such representations through its practice. It is construed through and by means of cultural configurations of images and interpretations of the "wild," the "tame," and the "domesticated" and through ideas of appropriate relations between particular categories of animals and the relationships between these and humans in particular rural spaces. Intimately linked with these are images and ideas of legitimate and illegitimate killings and killers in the English countryside. This essay aims to provide an anthropological ethnographic interpretation of some of the key representations of the particular animals involved in the practice of foxhunting and how the living, embodied animals which are figured in such representations are drawn into an engagement that produces an event which makes cultural sense to the people who participate in it.

Although there is a vast anthropological literature about hunting in those societies and cultures often defined as "hunting and gathering," little attention has been paid to other hunting practices in the modern world—particularly when those might be defined as sporting events. Perhaps this is because they have been seen as trivial and insignificant events unworthy of elevated academic attention; perhaps because it has been thought that they can easily be explained away as anachronistic; or perhaps, more likely, because they have been regarded as morally unacceptable practices.[1] However, those interested in human-animal relationships should consider these practices, because they involve complex sets of images and representations of animals and the natural world and complex structures of engagements with those animals and that world.

Foxhunting is a highly formal event. It is tightly structured and governed by written regulations and is constructed around, and bound by, strict notions of etiquette, appropriate behavior, and performance that allow for only certain forms of engagement between the hunter and the hunted. Foxhunting is also clearly marked as a special public event and has those "alerting qualities" and "conspicuous kinds of behaviour" framed by "imperative rules" which Gilbert Lewis suggests as indicators of ritual practice (19–22). He points to the

> peculiar fixity of ritual, that it is bound by rules which govern the order and sequence of performance. These are clear and explicit to the people who perform it. It is a form of custom. The fixity, the public attention, the colour and excitement or solemnity that go with such performance are what catches the anthropologist's attention. (7)

For Lewis, the other elements that make up this alerting quality include pageantry and ceremony, regulation and formality, direction by specialists, codes of dressing or costumes, often a complex lexicon and language not easily understood by outsiders, music or other sounds, and references to tradition.

Foxhunting has all of these elements. Not only do rules govern the hunting itself, but those who participate may only do so if they dress according to the codes set out for the event in general and according to the specific traditions of the particular Hunt.[2] This costume is a marker which sets the participants apart from the everyday world. An elaborate language code that is fully understood only by insiders provides terms for and descriptions of the animals and a lexicon of terms relating to hunting practice. Special use is made of music, formal gestures, and modes of address. The act of hunting is conducted by a specialist who is regarded as having a unique store of knowledge and an almost mysterious relationship with the animals with which he is engaged. From an anthropological perspective, then, foxhunting can be interpreted as a highly formal and complex ritual event enacted in English rural space.

Oscar Wilde memorably castigated those who participated in English foxhunting as "the unspeakable in full pursuit of the uneatable."[3] This witty and barbed social attack has remained a popular condemnation of the practice, but as an analysis of the hunt as an event it misses a crucial point—that the human participants do not pursue foxes. In foxhunting, humans are not directly engaged in hunting at all; they *are* involved, however, in a complex event that has at its center a culturally contrived predator-prey relationship between hounds and foxes. The aim of this essay is to suggest an interpretation of the event beyond the immediacy of the relationship of the killing of one animal by another. Although foxes are killed in this event, that is not its central concern—the concern is with the way these deaths are brought about and the meanings that they have, for they are very different from the deaths brought about by shooting, snaring, poisoning, and other forms of animal killing.

English Foxhunting: A Short Description

Foxhunting takes place during a season from late autumn until early spring. There are no fixed dates to this season. Hunting instead depends on when the farmers have finished harvesting their land in the autumn and when they begin planting again in the spring. There are some 220 registered "Hunts" that pursue foxes in England and Wales and each of these Hunts will hunt at least twice a week during the season. Each Hunt has several officials (see Buxton; *Baily's Hunting Companion*; and *Baily's Hunting Directory, 1999–2000* for detailed accounts of foxhunting's officials, organizers, and participants[4]), but this essay is concerned only with the Huntsman, who is responsible for working with the hounds and the practice of hunting. This figure is marked by his uniform, the key feature of which is his rich red jacket.[5] Apart from the Huntsman and his assistants (the Whippers In), those who participate in the event are mounted riders (the Field) and those who follow in vehicles or on foot.

On a hunting day the mounted and foot participants, officials, Masters, Huntsman, and hounds come together at midmorning at a prearranged point, which is often a country house, village green, local farm, or public house. This event, known simply as the Meet, is one of conscious pageantry—the Hunt is on public display. The riders are attired in the equestrian clothing which is deemed traditional to, and appropriate for, foxhunting; the horses are perfectly groomed; and the pack of pedigree hounds, the pride of each Hunt, are on show close to the horse of the Huntsman. The individual who hosts the Meet will serve drinks and light snacks. This is a social occasion for all who participate. Many local people will also come to see the Meet, to observe and celebrate the pageantry of the occasion and, if they know people in the Hunt, to engage with it as a social event in itself. Many, for example, will come to chat with the riders and foot followers and to allow their children to pet the hounds.

The Meet will last only a short time—thirty minutes or so—before the Huntsman calls for the hounds to be drawn together (he will have anywhere between fifteen and thirty-plus hounds with him), blows his horn, and sets off to the first place he plans to hunt. At this point different groups of people are significantly divided in their positioning and participation. A physical distance is maintained between the Huntsman and the mounted riders known as the Field. The mounted riders must follow the directions of the Field Master, the official for the day who will guide them across the countryside. Riders are not permitted to decide on their own route. The Huntsman will approach the first field, wood, or hedgerow—known in the language of foxhunting as a "covert"—and hunting will begin. The hounds are encouraged into the covert to try to find the scent of a fox. If there is a fox in the covert, it will probably flee quickly. Often the scent of a fox is present but the creature is long gone. Foxhounds hunt by scent; they do not look for foxes. While this is going on, the riders are held up at some distance and foot followers are expected not to approach too close. If the hounds do not find a scent the Huntsman will move them on to the next covert. If a hound or several hounds working closely together find a scent, they will communicate this by the tone and quality of their cries to the rest of the pack, which will quickly come to them. A fox may be put to flight and the hounds may still take a while to pick up the scent, or they may pick up the scent of a fox that left some time ago. From this moment, even though the fox is not present, hunting is taking place—they are in pursuit, not of a particular animal but of the scent of a particular animal. If the scent is strong (which depends as much on local climatic conditions as on how recently the fox was present), the hounds may begin to move at speed. If the scent is old, faint, or quickly evaporating, they will move slowly and hesitantly. The Huntsman will follow, often blowing on his horn to communicate what is happening to the other participants and to add a musical element to the general texture of the event. The mounted riders will follow some distance behind and may be allowed to jump hedges, gates, and other obstacles—a key element in hunting for those who participate for the thrill of an exciting equestrian challenge.

Many, though, have come to witness and enjoy the central contest that is now being enacted—the relationship between the fox (which is often not visible) and the hounds. If a fox becomes alert to what is happening, it will often take evasive action. It will attempt to disguise its presence in the landscape by taking a complex route across the countryside, which often involves doubling back or criss-crossing its own path. It may run among livestock or along a stream or a road to hide its scent. It may attempt to gain the safety of another animal's underground shelter. What is of primary interest to the human participants is how the hounds work as a pack, how they attempt to keep to the scent, how they communicate with one another, how they display their strength and agility as they move across the countryside, and how they try to resolve the problem if they lose the scent. All of this is judged in terms of aesthetics and efficacy. This is the hounds' performance. The Huntsman too is judged on his performance— his demonstration of his ability to understand and read his hounds, to understand the fox, to interpret the landscape, and to maintain the rhythm of the day's hunting. How he balances the need to let the hounds work on their own and to encourage and help them if they experience difficulties is fundamental. He will also be judged on his understanding, or lack of it, of the situations as they develop and his control, or lack of it, of the hounds.

Once the hounds have successfully started to follow the scent of a fox there are several possible outcomes: the fox may evade them and escape completely; the hounds may finally draw close enough to it to be able to see it and push on faster so that they catch and kill it; or the fox may take refuge in some sort of shelter which is discovered by the hounds. In the first case, when the hounds have lost the scent of the pursued fox, the Huntsman will encourage them to search in different directions around the point where they lost it (here his skill in understanding fox behavior comes to the fore), but if this is unsuccessful he will call his hounds to him and begin the event all over again. If the hounds succeed in catching their fox they will quickly kill it and the whole pack will surge around, tearing at the carcass. The Huntsman, when he reaches them, will usually dismount and blow "The Kill" on his horn. The mounted Field, if they have not been separated because of the difficulties of the terrain, will arrive but have no role to play here—indeed many will place themselves at a distance from or turn their backs on what the hounds are doing. Once the Huntsman has decided that the hounds have had enough satisfaction from tearing at the carcass he will call them to him and they will start hunting again. If the hunted fox takes refuge underground and the hounds scrabble at the point where they lost it, another procedure begins. The hounds are removed and a team of specialists with small terriers are brought in. All exits from the refuge are sealed and a terrier wearing an electronic collar is encouraged to go underground. It will pursue the fox until it can go no further. The Terrier Man registers this using an electronic receiver and will then dig down to the fox and terrier, remove the terrier, and shoot the fox with a pistol. The carcass will often be given to the hounds. Once again, hunting will resume when this has been completed.

Hunting will continue in this manner until late into the afternoon. The mounted riders may leave individually at different times and, when the Master and the Huntsman judge that it is time to bring hunting to a close for the day, the Huntsman and his assistants will bring the pack together, guide them back to their transport vehicle, and return them to the Hunt kennels.[6]

The Fox: An Ambiguous Quarry

England has a long tradition of construing the hunting of many animal species as a sport. Until they became extinct in England, boar, wolves, and bear were hunted by the aristocracy and other elite landowners. They also hunted deer of various kinds and hare, both of which are still quarry species. Informal foxhunting has a long history, but foxes were not perceived as a quarry species worthy of the aristocracy or landed gentry (see Ridley 2 and Carr 26–30).

It was only in the eighteenth century that the status of the fox was raised to one which made it worthy of formal hunting. The reasons for this are complex, but one of the central factors was the decline in numbers of the preferred quarry: deer, particularly Red Deer (Carr 22–24). In the mid eighteenth century the owners of packs of hunting hounds discovered that the fox, because of its speed and tactics when pursued, offered the possibility of "good sport," and new principles and methods of foxhunting were developed (see Beckford). There was a dilemma, however. For centuries the fox had been chiefly represented in negative terms and images as "vermin." It killed poultry and other domestic livestock belonging to humans, and when caught, whether deliberately sought out or simply discovered, it was unceremoniously killed as an undesirable animal. Heavily burdened with this image, and suffering "punishment" for it, the fox now became a substitute for the traditional, "noble," quarry species. As one anthropologist comments ironically,

> The fox appears not as the star of a ritual drama specifically written to his personal qualities, but rather as a second-rate substitute, an understudy drafted into the central role after the lead came down with a lingering terminal illness. While only the fox can lead riders on a chase suitable as a ritual of nobility, the riders have not been keen to transfer the fox's qualities to themselves. (Howe 295)

Hunting, as an aristocratic and gentlemanly pastime, had ennobled the animals that were central to it as its quarry. The attention paid to them through hunting gave them a special quality. Those who began to hunt the fox had to overcome what were regarded as the animal's far-from-noble "personal qualities" and to reinscribe and reimagine it. They managed to do so, and the fox did indeed become the "star" of a drama rewritten for it. It was still perceived by many as a rural pest, but it came to have a quasi-sacred status among those who hunted, and it was regarded as shameful, immoral, and unnatural to kill a fox in any way other than hunting with hounds. Although farmers and villagers might kill a fox for the practical reason that it was taking their livestock, this was not the acceptable behavior of a gentleman. Among those who hunted this was elevated to a crime, an act so heinous that it was named "vulpicide": the illicit killing, the murder, of a fox.

The idea that the fox was a pest persisted, but closer attention was paid to a widely held set of characterizations which were privileged in the imaginations of those who hunted. Central to this was a set of popular images of the fox as clever, wily, shrewd, immoral, and cunning—a trickster. These images could also be found in literary works such as *Aesop's Fables* and the medieval beast epics (particularly in the stories of the *Roman de Renart*: see, for example, Terry; Varty; and Varty, ed.). The newly developing form of foxhunting depended on these supposedly natural characteristics of the fox, which were recast culturally. It was not that the fox presented an interesting challenge simply because it could outrun hounds—the hunt was not a race—or because it was a dangerous animal which presented a risk when hunted, but because humans and hounds had to pit their wits and skills against what was considered to be a clever opponent. As foxhunting gained in popularity the event itself became the source of representations of the fox. These representations and the idea of "foxness" were reinforced and enriched through the continuous human experience of the animal when hunted. Such experiences were retold as stories of how foxes were found, lost, or killed; of the apparent strategies they used to disguise their presence and evade or confuse the hounds; and of the skills needed by the Huntsman and the hounds to effectively respond to these strategies.

Foxhunting in England today depends on, is constructed from, and continues to reinforce all of these elements, but in order to develop an anthropological understanding of its internal cultural sense it is necessary to explore more fully one of the characterizations of the fox mentioned above—that of its "immorality." The immorality of the fox is expressed in terms of its being a "thief" in its choice of food—particularly game birds and domestic livestock and poultry. This is perhaps an unusual characterization for a wild animal which is, on one level, simply acting out its animal nature. Other wild predators are not characterized in this way—they prey on other wild creatures, which is proper, acceptable, and appropriate. This is what the fox should do, but it "chooses" not to restrict itself to what are perceived to be the "proper" sources of food for a wild creature.

The fox is problematic because it will prey on animals which belong to humans, and for that reason its behavior and character become subject to cultural elaboration. The notion of "thief" is key here, for it expresses a relationship with the human world. The fox comes out of the world of the wild to intrude into the human world and into human concerns. But the perception of this act of theft is nuanced. The fox is the animal analogue of the human poacher (and the term "poacher" is often directly applied to the fox)—a rural thief for whom there is grudging respect. The poacher is very different in the popular rural imagination from the urban burglar, shoplifter, robber, or mugger. Rather than a dangerous and evil character, he is regarded as a rogue—someone who does something wrong but who is nonetheless admired, because what he does is not regarded as simple theft or anti-social behavior. In large part this admiration is associated with the animals taken by poachers. Rather than stealing domestic livestock, poachers concentrate on animals classified as "game"—for example, deer, pheasant and other game birds, trout and salmon—all of which have an ambiguous status.[7] They are wild animals, but wild animals over which someone asserts ownership. They are bred and raised, or at least protected, so that they are available in other sporting events. The poacher contests the legitimacy of ownership of these animals and subverts this relationship between the protected animals and their owners by treating them as wild creatures which can be simply killed for food or for sale and by killing them in a non-sporting, non-public way, usually under the cover of darkness. Both the human poacher and the fox are problematic because they intrude, in unacceptable ways, into other human-animal relationships. But they are poachers for different reasons. Whereas human poaching is the inappropriate taking and killing of certain wild/game animals, the terms are reversed for the fox. The fox only becomes a poacher when it kills domestic animals.

Some in the countryside will simply kill foxes as pests. Gamekeepers and farmers will shoot, trap, and poison foxes because they feel that it is necessary to control their numbers, but those who hunt still regard this as an inappropriate form of death. The special status of the fox among those who participate in and support hunting is marked by the strong sense that humans should not directly kill foxes at all—foxes not only should be, but deserve (in the positive sense of "merit") to be, hunted and killed by hounds. The fox may be a pest, but it is a creature worthy of respect, and this respect should be expressed through making it the central character in what is perceived as the noble art of hunting. Although it might be characterized as "vermin," and more often by the less emotive term "pest," the fox is, as it were, a higher class of vermin than other animals. The language and processes of vermin control are those of elimination, eradication, extermination, and destruction. The fox, however, should not be subject to such treatment, and a central perception of, and justification for, hunting is that it is a "natural" relationship between predator and prey in which the natural predator is turned into prey. The fox should not be treated as a mere criminal animal, but should be engaged within an event akin to a trial in which it can express itself, reveal its qualities, and, in a sense, prove itself.

Members of other pest species are not individuated. No relationships are formed, for example, between individual rats and their human killers—they are merely destroyed as members of that species. This is not so with the fox, and the species should not be the subject of indiscriminate slaughter. Although foxes are not known as individuals when they are in the wild, the foxhunt is constructed in such a way as to bring about a relationship with only one fox at any one time. Although no individual fox is selected for hunting, from the moment a scent is found it is that particular fox which is hunted, and it is responded to as a unique individual with unique characteristics. Once the hounds become engaged with the scent of a fox, it is regarded as inappropriate and "bad form" for them to be allowed to leave that and set off after a fresh fox if the opportunity presents itself.

The Foxhound: A Cultured Hunter

Xenophon, in the opening sentence of his treatise *On Hunting*, makes the highest claims for the special status of hunting hounds when he writes, "Game and hounds are the invention of the gods, of Apollo and Artemis" (367). A variety of other historical records, literary sources, and visual representations from different cultures and societies, although less elevated in their claims, nevertheless show that canines classed as hunting hounds have occupied a privileged position in the households of monarchs, aristocrats, nobility, and the landed gentry (see, for example, Cummins; Gilby; Longrigg; Phoebus). Accorded special status, they have been highly valued and prized for their role in hunting. The foxhound in modern England continues to have a similar status. In the language of hunting (deer and hare hunting as well as foxhunting) these animals are never "dogs" (unless one is referring specifically to the males); they are always "hounds"—a distinction which is clearly marked, in the context of hunting, by referring to all other domestic canines, however illustrious their pedigree, as "cur dogs."

Since it was first used for foxhunting, this type of hound has been specially bred for the purpose and is the subject of enormous cultural elaboration. Each Hunt has its own pack of hounds and the pedigree of some of these lines can be traced back through the records to the eighteenth and nineteenth centuries. Although those interested in hunting often speak of a pack of hounds acting as though they were an equivalent to a pack of wild canines, this is far from the case. These animals are a cultural creation based on a natural form. They are the result of highly selective breeding based on the skills, choices, decisions, and imagination of those who control the process.

Foxhounds are bred for a variety of qualities. The physical, anatomical quality is probably the most easily controlled, influenced, and shaped. Foxhounds have to run perhaps fifty to sixty miles in a day, and they must have the build to be able to do this once or twice a week during a long season. The Huntsman (or whoever is responsible for the breeding) must think about maintaining a particular size, athletic build, and robustness, for these are working animals. Crucially, though, the foxhound is an aesthetic product. The different pedigree lines and the individual hounds of those lines are the result of human imagination, thought, will, and desire. Breeders seek to maintain or develop a particular "type" of hound in their kennels. "Type" refers not only to the physical shape and size of the animal—its conformation—but also to the type and color marking of its coat.

Each breeder will also be searching to improve less tangible qualities. Foxhounds are "scent hounds" rather than "gaze hounds." They do not seek their prey by looking for it but by searching for its scent. Different hounds have a greater or lesser ability in this. All must have a basic ability to follow a scent when it is there, but some are regarded as particularly skilled when the scenting conditions are poor. Such individuals are valued in the pack because they can keep a weak scent in their nostrils or are able to follow it when the fox has attempted to disguise the scent on a road or by running through fields and enclosures where there are other animals. The Huntsman learns to trust these hounds and to rely on them when the conditions are poor—he would always like to have more of these in the pack.

The "voice" of the hound is another such intangible quality and one that is intimately connected with scenting ability. Unlike domestic (pet) dogs, hounds do not simply bark; they also have a particular set of cries which they use when hunting. As soon as a hound picks up on the scent of the fox, it should signal this to other members of the pack through its cries—a hound must be able to communicate. As more and more hounds find the scent the cries intensify and are interpreted by the human participants as a commentary on the developing relationship between the hounds and the fox. This cry is not merely a utilitarian signaling device; it is an important aesthetic element, and breeders hope to produce a complex set of voices and harmonious sound within the pack. Hounds are regarded as having bass, tenor, alto, or soprano voices, and these should be well-represented. A mute hound—one which is unwilling or unable to use this quality of canine sound—is incomplete. The collective sound of the pack when hunting is referred to as *speaking*, and this sound is both appreciated and responded to aesthetically. Not only should the hounds' cries have content which communicates a message, but the sound, purely as sound, should have a quality which is registered by referring to it as "the music of hounds." A simple set of canine sounds are transformed in the imagination of the human listener—the natural becomes cultural.

Although foxhounds as a breed are used in other forms of hunting (deer, hare, and mink), those which belong to a foxhunting pack should respond only to the scent of a fox. At times, when hunting, they will lose the scent; they will come to points where the fox has apparently attempted to confuse them or to disguise its presence. It is at those moments that the Huntsman needs hounds that possess what is referred to as "fox sense." Once again, this is not a quality which can be immediately bred for, and it is highly valued in the hounds that have it. "Fox sense" is spoken of as a hound's ability to think like a fox, to be aware, as it were, of its own intentions and simultaneously to be able to project itself into the mind of the fox as it apparently plans and develops tactics of avoidance. This is not quite regarded as a natural instinct—it is an attribution of a form of understanding to the hound, a sense that it is aware of what is expected of it and that it can project this into the future. Here there is a sense that one animal can understand and interpret the mind and behavior of another. The attribution of "fox sense" is both anthropomorphic and, at the same time, an animal equivalent of anthropomorphism. This "fox sense," though, should only operate in particular ways. It is not expected and not desirable for a hound to plan ahead and catch the fox in any way other than by following the scent. It is not expected that a hound should be able to take a different route across the countryside in an attempt to cut the fox off or to ambush it in some way. The qualities of "fox sense" are captured in the comments of two hunting commentators. Sir Newton Rycroft, in response to a question about this, argued, "They should have sense but they should not be thinking. If they go where they think the fox has gone they are absolutely useless."[8] Good "fox sense" should simply lead a hound to find the scent again once it has been lost. As Lord Mancroft, a former Master of Foxhounds, commented,

> Foremost among the attributes that separate great hounds from good ones is "fox sense." Some hounds seem to know where the fox has gone, for no apparent logical reason. At a check [the point where the hounds have lost the scent] one hound may work away by itself, as if it knows something the others do not. Huntsmen learn to trust these hounds. (90)

Hounds should resolve the problem of finding and catching the fox only according to the terms set out by the fox. This means following only where it has gone. Foxes are thought to attempt to outwit the hounds, but the hounds are not expected to engage in such a practice. Once a fox is being pursued it is referred to as the "pilot": it leads. The course and the difficulties should be set by the fox and resolved by the hounds.

Whatever the qualities and skills of individual hounds, they must work as a pack. Huntsmen, like the directors of many sports, want to have performers who will work as a team rather than a disparate group of stars which the rest follow. They must all work equally at the task, and it is their performance as a pack that is judged. The working of the pack is the central feature of hunting as a performance which many participants come to watch. The pack is judged by how effective it is in finding a scent, especially when the conditions are difficult. Those watching will comment on whether the hounds are determined or lazy, intelligent or foolish, enthusiastic or uninterested, listless or alert. Once one or several hounds have found a scent they are expected to communicate this excitedly to the others. A hound that finds a scent and silently follows it on his or her own is failing in its role as a pack member. Hounds should not hunt on their own—this is a collective enterprise. Equally, a hound which "speaks" too quickly, when it has not really found a scent, is failing. Significantly, the term for this is "babbling"—the same word used for humans who speak without purpose or without making sense.

Once the scent of a fox has been found the hounds will be judged by how well they hold it as a pack and their style in following the line of the fox across the countryside. Hounds will often "check"—come to a halt because they have lost the scent or because they are confused. Sometimes they will turn to the Huntsman for help, and he encourages them to search in the direction he thinks the fox might have gone. Once again hounds are judged and appreciated for the determination and intelligence they show in the way they try to solve the problem for themselves and regain the scent.

Although there is no plan to hunt any individual fox, once the hounds pick up the scent of a fox, it is expected that they will follow that scent and not switch to the scent of another if they come across it. From the moment of scenting the hunt becomes that of one particular fox and much of the interest comes from the development of the relationship between this individual and its actions and the hounds. If the hounds have been successful in following the scent, then at some point they will actually lift their heads and see their prey. They should increase their pace, surge forward, and kill it. Significantly there is no commentary and no judgment made on how the hounds actually kill the fox. If the fox evades and escapes them in these final moments, comments certainly will be made about their failure, but the death of the animal merely constitutes the culmination of hunting and a way of bringing the hunt to an end. The death is treated as a normal animal death brought about by an appropriate predator.

The following section will explore more closely the nature and structures of hunting, but it is worth commenting here on the perceptions and representations of the foxhounds as a hunting pack. Although breeders and Huntsmen often speak of a pack acting like, for example, a pack of wolves, there are some important differences. A pack of wild canines when hunting are doing so because they are following their natural instincts to obtain something significant—food. A pack of foxhounds is not working in exactly the same way. They certainly hunt in the sense that they attempt to find, follow, catch, and kill their prey—and this seems based on certain natural instincts—but they are doing so in ways that have been established by their human owners. The hounds are therefore hunting for the humans. In an important sense they are not hunting for their own purposes, although they seem purposeful in their pursuit; they seem to understand the immediate purpose from the repeated experience of going hunting, but again this understanding seems to be of what is required of them by their human masters. To return to Oscar Wilde once more, he was correct in his comment that those in pursuit were "in full pursuit of the uneatable," because hounds do not eat the fox that they kill, and fox does not constitute part of their diet. Hounds are neither capturing food for themselves nor are they doing so, as in many other hunting events, for humans. The purpose and meaning of hunting are the processes itself. Although directed toward an end, that end has no significance of the sort it has in other hunting, which is marked by obtaining meat, skins, furs, or trophies. As was mentioned above, the death of the fox is marked, but it is not celebrated. The fact that it has occurred is important and the manner in which it has been brought about is crucial, but the actual obtaining of a dead fox has no significance other than to register a successful hunt and its appropriate conclusion.[9] The dead animal is not individuated and incorporated into the human world as a stuffed trophy,[10] and the human participants do not pose for photographs with the dead animal, as is the practice in many hunting events, because none of them can claim to have brought about the death of the animal through skill and prowess—none of them have hunted it and the kill is hence not an immediate human achievement. It is a human achievement, but at one remove. Judgments about success or failure in hunting are expressed through a commentary on the hounds themselves, and also through a commentary on the Huntsman. This will remark on his skill, or lack of it, in breeding the hounds and how he has helped or hindered the hounds during a particular hunt. But even though the successful killing of a fox has been achieved, partly at least, through his close association with his hounds, it is the achievement of the animals that is celebrated, and as they tear at the carcass the Huntsman encourages the hounds with excited cries and praises them lavishly.

Hunting as Practice

This form of hunting, although it shares some of the characteristics of hunting for essential food, is a sport—a non-necessary, non-utilitarian, game-like event which is freely entered into. Although it forms a regular part of the lives of those who participate, it is set apart from their everyday lives. The only person for whom it forms a defining feature of everyday life is the professional Huntsman, for whom it is simultaneously his work and his sport. Although hunting in what are called "hunting and gathering societies" is usually embedded in a complex set of views and beliefs, often religious, about the appropriate relationships between humans and their prey, the animals that are the focus of this attention are usually given as few chances as possible to escape. The human hunters need to kill these animals for survival. In sports hunting events, though, humans must voluntarily reduce or restrict their ability to kill animals, and they must create challenges and difficulties for themselves in order to create sport. If, for example, deer were to be driven from a wood and forced into a fenced area in which they could be shot, this would not be sports hunting but simple slaughter. Sports hunting has rules—sometimes accepted practice, etiquette, and norms, sometimes codified as written regulations—which are integral to each example of it and which allow it to be hunting rather than any other killing of animals. It is the self-imposition and willing acceptance of difficulties that creates the challenge, contest, emotional engagement, excitement, and if they are successfully resolved, the sense of accomplishment.

Matt Cartmill's *A View to a Death in the Morning* is one of the few anthropological engagements with hunting as sport. In this rich and innovative work Cartmill's focus is on understanding hunting as symbolic behavior. His fundamental point is that

> hunting in the modern world is . . . intelligible only as symbolic behaviour, like a game or a religious ceremony, and the emotions that the hunt arouses can be understood only in symbolic terms. . . . Hunting is not just a matter of going out and killing any old animal; in fact very little animal-killing qualifies as hunting. A suc-

cessful hunt ends in the killing of an animal, but it must be a *special* sort of animal that is killed in a *specific* way for a *particular* reason. (29, emphasis added)

Other fundamental elements that Cartmill builds into his definition are that the quarry must be a wild rather than a domesticated animal, that hunting must involve physical violence to the animal, that this violence must be inflicted directly, and that the hunter's assault on the quarry must be premeditated (for example, a car driver's attempt to kill a pheasant or a rabbit on the road is not hunting, even if it is deliberate). This last point links closely with the views of Ingold, who also stresses that human hunting must be explored as social and cultural practices and argues that the difference between human hunting and predation is that "the essence of hunting lies in the prior intention that motivates the search for game," whereas that of predation is in "the behavioural events of pursuit and capture, sparked off by the presence, in the immediate environment, of a target animal or its signs" (*Appropriation* 91). In this sense the gamekeeper walking through the estate who sees a fox, rabbit, or crow and shoots it is not hunting; he is merely taking the opportunity to kill a pest. Hunting is thought about, planned, organized, and intentional; it is not the casual killing of wild animals that people happen to come across while they are engaged in some other social activity.

English foxhunting combines all these elements set out by Cartmill. People come together on a particular day in order to hunt; there may be other, micro social activities—such as discussions about business, plans for a Hunt activity, or a private dinner party—going on simultaneously, but the encompassing activity is hunting. It is an event that centers on the deliberate intention of finding, pursuing, and attempting to kill particular wild animals—foxes—during the course of a hunt. It should be emphasized that the hunters are only interested in foxes—this is foxhunting. As mentioned above, if the hounds put rabbits, hare, deer, or other animals to flight during the hunt they should not be pursued—something that would shift the activity toward, in Ingold's terms, predation—and a Huntsman will soon punish hounds that divert their attention in this way. The animal must be given a chance to escape; it should never in any way be held up, restrained, or enclosed so that the hounds can catch up with it and be certain of killing it. There is a slight difference from Cartmill's definition of the nature of the violence inflicted on the hunted animal. Although there is direct violence, rather than being administered by a weapon controlled by an individual human hunter, it is administered by means of other animals (the hounds) under the control of the Huntsman. It is through the use of hounds that the human participants voluntarily restrict their abilities to kill foxes—the hounds may or may not find the scent, they may or may not be able to follow it, and the possibility of direct engagement with the fox is always in doubt.

Once again this can be related to the apparent aims or ends of the event and to the comment made earlier that achieving the death of the fox is not the central feature of foxhunting. The focus of attention is on the development of a set of potential encounters, actual encounters, and failed encounters between the hounds and the fox that create excitement and dramatic tension for the human participants. Sometimes hounds, when entering a wood, will find and kill a fox immediately without its having the chance to escape—something referred to as "chopping." Although this is an efficient and effective killing of a fox by hounds, within the context of the hunt it is not regarded as a proper hunting event because there has been no contest, no tension in the development of the relationship between the two sets of animals, and hence no possibility of human engagement or the transmission of emotion to the human participants. Discovery and death, the beginning and the end, have been conflated into one moment rather than being tenuously linked through time and across the landscape. The death of the fox has no significance without the rhythms of the shifting relationships between the two sets of animals. Its death is sought after and desired but only as a resolution of the problems that foxhunting sets for itself.

The role of the hounds and their relationship with the human participants is a key element of this form of hunting. As was argued at the beginning of this essay, the human participants are not directly hunting—they do not attempt to track, find, pursue, or kill anything in the event. Foxhunting is structured around hounds doing these activities and human participation involves following, witnessing, and judging them.[11] The only person to come close to hunting is the Huntsman. He does attempt to find the fox, but he does not attempt this in any immediate and direct way. As a hunter he will think about where he might find his potential prey; he will read the signs of the countryside that might suggest its presence, but he then directs the attention of his hounds according to his insights and instincts. A Huntsman could successfully engage in hunting foxes without any of the riders behind him. Indeed some Huntsmen have commented that this is the "purest" form of hunting, because it removes the need for them to be concerned with any form of performance not immediately related to the needs and demands of the "science" of hunting itself.[12] Although this would be hunting, it would not be "foxhunting" in its totality as a cultural event. The intimate and interlinked engagement of Huntsman, hounds, and fox is the central core of a more complex public ceremonial and dramatic event.

It is around this core of engagement that the sporting quality of the event develops. The sport is essentially a contest between the fox, which attempts to remain absent, unengaged, distant, and alive, and the Huntsman/hounds who wish to force its presence, to engage with it, to bring it close, and ultimately to kill it. It is the manner of this enactment that is crucial, though, for it is this which makes it a sporting cultural event rather than a natural encounter of predator and prey. Humans set rules of engagement and etiquette, notions of fair play, and acceptable practice which frame the encounter between these two sets of animals. The fox, of course, is not cognizant of these rules and is merely following its natural instinct, perhaps combined with behaving in a way based on learned experience if, for example, it has been hunted before. This biological behavior, though, is reframed, in the minds of the observers, as a performance that is responded to with aesthetic judgments as well as appreciation of the dramatic situations it creates. The hounds too are perceived as performing. They may be partially behaving in accordance with natural instincts, but they are certainly responding to training and may, because of the experience of punishment and praise, have some understanding of what is expected of them and how they should act. Ideas about their motivations, intentions, and performance are highly elaborated and much discussed by Huntsmen and others interested in hounds. Foxhounds are ambiguously placed between the world of the wild and the world of humans. At times they are spoken about as having an instinct for hunting and a natural instinct to hunt as a pack. At others they are spoken of as having an understanding of what is required of them. They are spoken of as trying hard to please the Huntsman, as having a sense of responsibility, as being aware of the fact that they are special animals—foxhounds—of which much is expected.

Woven through foxhunting is a shifting balance between ideas of, and representations of, the natural and nature and culture, expressed as human concerns. This is not simply a natural relationship between animal predator and animal prey—although it has elements of that; neither is it a relationship between human hunter and animal prey—although it has elements of that too. The event is both alternately and simultaneously natural and artificial. Through the preparation, development, and resolution of hunting, human concerns about relationships with animals in rural spaces are enacted in an expressive manner. Underpinning English foxhunting are shifting patterns of animal representations. To return to an argument developed earlier, of all the wild animals in the countryside the fox is the focus of interest because of its ambiguous nature as a predator whose behavior impinges on the human world. A fox as a wild animal should, properly, as part of its natural behavior, hunt and kill other wild animals, but it will often kill domestic livestock for its purposes—animals that should be killed by humans for their purposes. This is perceived to be improper. The fox is not acting as a proper wild animal; it is not hunting—in the sense of tracking, pursuing, and capturing its prey—when it focuses its attention on poultry, lambs, or young game birds. These are controlled, contained, and have no chance of escape; this then is not hunting but raiding and hence improper. Hounds too are ambiguous; they are domesticated animals (but differentiated from pet dogs) and yet they hunt—something which domesticated animals do not normally do. Although living under the care and control of humans they are allowed, in the context of this event, to revert partially to the wild (again something which humans normally strive to avoid in domesticated species). They are encouraged to act in accordance with what are perceived to be their instincts (although as we have seen these are reshaped by humans), and they are physically unrestrained while doing this. Hounds are encouraged to hunt but they are not in search of a source of food; they are artificial predators in pursuit of an artificial prey and yet they are truly hunting. Those who hunt believe that it is proper for foxes to be hunted with hounds (indeed the death of a fox by any other means is at best a necessary evil and at worst, if caused for pleasure, an ignoble and demeaning death), and yet this comes about only because the fox is an improper hunter.

Shared Lives: Shared Worlds

Tim Ingold, in his critical rethinking of the relationships which humans have with animals as both their hunters and their domesticators, suggests that to conceive of an animal as being truly wild, as living an "authentically natural life," only if it is "untainted by human contact" is limiting ("From Trust" 10). Just as humans live their lives with animals in their worlds, so animals live with humans in theirs. Many societies, particularly in the West, are concerned about what those relationships ought to be, and there is a sense in many of the commentaries on this that there ought to be little direct engagement between the two. Humans may view "wild" animals but, in the main, they should avoid all contact in order to preserve what is perceived as some sort of authenticity in their wildness. For many people in present-day England, foxhunting is an unacceptable anachronism because of its openly intrusive engagement with the natural world and because it involves the attempt to bring about the death of a wild animal—something humans should have nothing to do with. For those who oppose foxhunting this death is immoral because it is gratuitous and fulfills no acceptable needs. But the participants view it very differently. Foxhunting is a celebration of a fully engaged life with animals and of a close experience with them; it places humans centrally in a web of relationships with animals, both wild and domesticated, in rural landscapes that are also landscapes of meaning and intense emotions.

From the perspective of this practice wild animals have a place in a shared, lived world with humans; they have experiences, often merely fleeting, of humans just as humans have experiences, often equally fleeting, of them. The representations of their lives are written around and through the web of relationships involving wild animals, domesticated animals, and humans. The world of animals has an immediacy and a set of significances for people in rural space, and wild animals have no neutral, non-cultural space in which to exist outside of these significances; relationships are interconnected.

These humans have a particular focus on foxes because of their own relationships with other animals—in this case a relationship constructed around ownership of them and ways of killing them. Those engaged in foxhunting are not interested in what the fox *is*—its being—so much as what it *does*—the nature of its engagement with the world. Although they do not have an immediate relationship with the fox apart from that of occasional sight in everyday life, they have one that is both predicated on, and mediated through, their relationships with other animals. Foxes, unlike wild animals in some places in the world, are not dangerous to humans. They do not directly threaten people, but they are dangerous and threatening to other animals in human care. Domesticated animals should be protected by humans; in part that is what the relationships of domestication involve. In this sense the fox is an enemy, but it is also a rival in that it seeks to kill the animals which humans should kill.

As a result of the relationships which the fox has with animals in its world, those who hunt have configured a set of relationships with other animals that requires these animals to live their lives in human-centered environments. Humans have developed relationships with hounds, for example, in order to have a more engaged relationship with the fox, which in turn means that hounds are tied into a set of relationships with fellow hounds, with humans, and with a particular wild animal, and each relationship assumes a greater or lesser priority at different times. The world and space of hounds is populated with wild animals and other domesticated animals, but relationships with them are forbidden. Although they have not been discussed in this essay, horses are, of course, the other animal that is the subject of enormous cultural elaboration and complex representation in this context. Riders have a close, immediate, and highly individual relationship with their horses, and through this they are able to form a relationship with both foxes and hounds and a relationship with, and experience of, the countryside. It is through the horse that the humans are able to participate, not simply as observers of the contest between fox and hounds, but as fully committed actors in the drama across its ever-changing setting.

As has been set out above, this contest is not simply one of a predator seeking its prey for its own ends; it is a challenge and a contest set up for human benefit, to express human concerns, and humans ought to be present during its enactment. Without their presence it loses its significance and has no meaning except for those animals immediately concerned. This event is not about animal concerns and it is not created for the benefit of animals. It is a drama of human concerns, and humans must be there to experience, witness, and draw significance from it.

All of these animals are enmeshed in a web of significance that has been devised for them by humans. From human thoughts, feelings, observations, beliefs, and imaginings about animals come representations of them. However "natural" they may be, animals will always be, when the object of human attention, "cultural" and will be drawn into forms of engagement with the human world. It would seem that humans must necessarily engage with, or distance themselves from, animals in human terms, according to social and cultural representations of them, rather than according to what animals "are" or might "be," because it is hard to imagine how they might understand and respond to any definitions which, for example, foxes, hounds, and horses might have of themselves. What those representations are for some and what they ought to be for others is part of the contested story of what it means to be human.

Notes

1. For a recent sustained academic engagement with the practices and the social and cultural meanings of hunting with hounds in England, see Norton, "Place of Hunting."

2. Throughout this essay the use of the word Hunt, with a capital letter, refers to

the foxhunt as the social entity, or association, rather than to the activities that constitute hunting itself.

3. This line is so often misattributed in Wilde's work that it is worth giving the full context. The comment appears in Act I of *A Woman of No Importance* (first produced on 19 April 1893 and first published in 1894).

> Lady Caroline: I am not at all in favour of amusements for the poor, Jane. Blankets and coals are sufficient. There is too much love of pleasure amongst the upper classes as it is. Health is what we want in modern life. The tone is not healthy, not healthy at all.
> Kelvil: You are quite right, Lady Caroline.
> Lady Caroline: I believe I am usually right.
> Mrs. Allonby: Horrid word, "health."
> Lord Illingworth: Silliest word in our language, and one knows so well the popular idea of health. The English country gentleman galloping after a fox—the unspeakable in full pursuit of the uneatable.

4. Although officials other than the Huntsman are not considered in this essay, it is important to note that the Master of Foxhounds, who can be male or female, is the most important public figure in any Hunt, because he or she is responsible for its organization. There may be more than one Master of any Hunt, and the Masters will hold the role collectively. However, on a hunting day only one will exercise this authority. The Master is in charge of hunting on any particular day, although in practice the decisions about where to hunt are usually made in consultation with the Huntsman. He is free to decide *how* to hunt.

5. Dress codes are actually more complex than this. Other officials and specially invited members will also wear red coats, and a few Hunts even have different colored coats for the Huntsman and officials.

6. What is offered here is a highly simplified and schematic outline of a day's hunting. There are all sorts of variations, in response to situations as they develop, which make any hunting day more complex than this.

7. This poaching is perceived very differently than is the illegal taking of, for example, sheep, cattle, and horses, a less nuanced form of theft and more akin to the American notion of "rustling."

8. In an interview with the author.

9. Once again, the issue is slightly more complex than allowed here. In the present political rhetoric in England about the acceptability of hunting with hounds, it is an important part of the case presented by the pro-hunting lobby that foxes are a pest, that they must be culled, and that foxhunting is an efficient method of doing this. The fact that they are able to kill foxes in this manner shows, in their terms, that they are able to function as part of an efficient pest control operation.

10. The heads, tails, and lower part of the feet are sometimes cut from dead foxes and mounted by taxidermists, but these are not really trophies. They might be taken or given to someone to commemorate a special occasion—a hunt that happened on a particular birthday, a last hunt before retiring from an official role, et cetera—but the stuffed remains function more as a souvenir, as a marker of memory, than as a celebratory trophy.

11. Although it has been mentioned in passing, what has not been elaborated on

156 *Garry Marvin*

is the fact that foxhunting is also an equestrian event with many people participating for the enjoyment of exhilarating galloping and jumping in parts of the countryside to which they do not normally have access.

12.   A Huntsman understands that he must produce "a good day's sport" for those who have paid to participate as members of the Field. He will certainly feel pressure to move on perhaps more quickly than he would like if, for example, it does not appear that there is much scent in a wood and the riders are getting bored waiting for him and the hounds. To fulfill his responsibility to the Field, he will often work hard to create situations that allow for exciting riding.

Bibliography

Baily's Hunting Companion. Cambridge: Baily's/Pearson, 1994.

Baily's Hunting Directory, 1999–2000. Cambridge: Baily's/Pearson, 1999.

Beckford, Peter. *Thoughts on Hunting.* 1781. London: J. A. Allen, 1993.

Buxton, Meriel. *The World of Hunting.* London: Sportsman's, 1991.

Carr, Raymond. *English Foxhunting.* London: Weidenfeld and Nicolson, 1986.

Cartmill, Matt. *A View to a Death in the Morning: Hunting and Nature through History.* Cambridge, Mass.: Harvard University Press, 1993.

Cummins, John. *The Hound and the Hawk: The Art of Medieval Hunting.* London: Weidenfeld and Nicolson, 1988.

Gilby, Walter. *Hounds in Old Days.* 2nd ed. Revised and expanded by C. M. F. Scott. Hindhead: Spur, 1979.

Howe, James. "Foxhunting as Ritual." *American Ethnologist* 18.2 (1981): 278–300.

Ingold, Tim. *The Appropriation of Nature: Essays on Human Ecology and Social Relations.* Manchester: Manchester University Press, 1986.

———. "From Trust to Domination: An Alternative History of Human-Animal Relations." *Animals and Human Society: Changing Perspectives.* Ed. Aubrey Manning and James Serpell. London: Routledge, 1994. 1–22.

———. "Hunting and Gathering as Ways of Perceiving the Environment." *Redefining Nature: Ecology, Culture, and Domestication.* Ed. Roy Ellen and Katsuyoshi Fukui. Oxford: Berg, 1996. 117–55.

Lewis, Gilbert. *Day of Shining Red: An Essay on Understanding Ritual.* Cambridge: Cambridge University Press, 1980.

Longrigg, Roger. *The English Squire and His Sport.* London: Michael Joseph, 1977.

Mancroft, Lord. "Now's the Time to Spot Talent in Other Kennels." *Hunting with Country Illustrated* 7 (Feb. 2000): 89–90.

Norton, Andrew. "The Place of Hunting in Country Life." Ph.D. diss., University of Bristol, 1999.

Phoebus, Gaston. *The Hunting Book.* London: Regent, 1978.

Ridley, Jane. *Fox Hunting.* London: Collins, 1990.

Terry, Patricia. *Reynard the Fox.* Berkeley and Los Angeles: University of California Press, 1992.

Varty, Kenneth. *Reynard the Fox: A Study of the Fox in Medieval English Art.* Leicester: Leicester University Press, 1967.

———, ed. *Reynard the Fox: Social Engagement and Cultural Metamorphoses in the Beast Epic from the Middle Ages to the Present.* Oxford: Berghahn, 2000.

Wilde, Oscar. *A Woman of No Importance. Complete Works of Oscar Wilde.* London: Book Club, 1976. 431–81.

Xenophon. "On Hunting." *Xenophon: Scripta Minora.* Ed. E. C. Marchant. Cambridge, Mass.: Harvard University Press, 1968. 367–457.

9 Displaying Death, Animating Life: Changing Fictions of "Liveness" from Taxidermy to Animatronics

Jane Desmond

In October of 1999, TNT released its widely ballyhooed feature *Animal Farm*, based on the novel by George Orwell. Several earlier attempts had been made to bring the novel to the screen, but it was not until the recent development of animatronics (the use of robotics, computer-generated animation, and live puppeteers to make constructed animals move) that a convincing adaptation became possible. Other recently released features, from *Babe* to *Jurassic Park*, and a plethora of new commercials (such as talking frogs selling Budweiser beer) have also drawn on the breakthrough technology of animatronics, as have museum displays and theme parks.

In this essay, I argue that this "breakthrough" is not really a breakthrough at all, but merely the current version of a historically significant yet underexamined phenomenon—the creation of "liveness" in animal display. This human passion to create, re-create, and animate three-dimensional animal bodies intrigues me. It stretches back in its current forms in the U.S. at least to the nineteenth century, when taxidermy became widely practiced. I suggest that low-tech taxidermy and high-tech animatronics have more in common than is first apparent. Both are intensely ironic practices and call for a compelling intimacy between human bodies and animal ones.

In taxidermy, humans kill animals and then manipulate their dead bodies to look alive. In animatronics, humans build fake animal bodies, get inside them, and, through their own bodily motions, "bring them to life." In order to better understand these passions to create fictions of liveness, I consider first of all the facticity of death, that is, our treatment of dead bodies. To reveal the specificity of our treatment of animals, I contrast the after-death treatment of animal and human bodies and compare the conventions of their public display.

In the first part of this essay, I examine the actual processes through which dead animal bodies are manipulated taxidermically, and how these have changed over time in pursuit of ever-heightened criteria of "realism." I contrast these processes with selected aspects of the display of dead human bodies. Finally, I analyze the technical processes of designing animatronic bodies, a form of taxi-

dermy that does not require death to bring the body to life. I articulate the ironies on which all these displays depend for their acceptability and power, and
analyze the processes through which these ironies are masked and naturalized.
Cases where the acts of display contradict or blur the social and conceptual di-
viding line between human and non-human animals are especially important.[1]
Throughout this analysis I hope to reveal more fully the roles animals and their
bodies play as our defining interlocutors even when they are dead or created
solely by human hands.

Taxidermy

The sharpest irony to traditional taxidermy is, of course, that it requires
the death of the animal in order to resurrect it as nearly as possible to a "lifelike"
state. To meet taxidermy's goal to "capture and preserve the vitality and living
energy of the animal in its natural state," the animal must first be killed, and
then all marks of killing must be erased (Laughing Elk Studio). Bullet holes
must be excised, their round entry points sliced horizontally and stitched into
invisibility. Early taxidermy manuals give advice on how to do this, and also
advise suffocating wounded animals rather than inflicting more gunshot wounds
when mounting is planned.

The death allows for stasis—live animals move, and move in ways we cannot
always control. But this stasis must also be reanimated. Taxidermic aesthetics
have developed from static displays (a duck standing on a wooden plank) to
arrested action poses, to the environmental dioramas first pioneered in the early
parts of the twentieth century. These last animate the animal body in time by
situating it in an implied narrative that includes moments "before" and "after"
that which we see frozen in front of us.

Classical poses, that is, "typical" arrested-motion stances, became associated
with different species of animals. For example, an ermine in a classical pose
would be standing upright, face slightly turned to one side as if looking at something in the distance, mouth open, short arms raised and reaching slightly forward. These arrested-motion poses captured, or rather indexed, the now lost
capacity of the (once living) animal to move, freezing the leaping tiger mid-
pounce, like a snapshot capturing motion forever in midair.

The individual death is masked not only through the theatrical staging techniques that "bring the animal to life" (lighting, fake foliage, painted backdrops)
in museum settings, but also in the transmutation of that individual animal into
an "example" or a "specimen" standing in for a whole species. By contrast, in
trophy displays (the deer head mounted on a plaque in the den, for example)
the fact of death is nearer—the act of killing is indexed as the mark of hunting
prowess, after all—but still the lifelikeness is indicated through the particularity
of the pose of the head, the care with which nostrils are painted practically
quivering, and the deep pools of dark glass eyes that glint in the light. Taxidermy
supply houses offer a full range of posed "mannekins," glass eyes, ear forms,
and mounting bases to choose from. The taxidermist can specify a full upright

or semi-upright stance, head turning to the right or left, shoulders straight-on or offset with a dramatic sweeping turn of the uplifted head. Sales catalogs promise that these forms "just come alive" when skin is mounted on them (McKenzie Taxidermy Supply). Presumably these invocations of arrested motion—the alert pose at the edge of a clearing, nostrils flaring to catch the hunter's scent, ears cocked for the betraying twig snap—index the final moment of life, as the hunter saw his prey. The implied "naturalism" of the poses situates the animals in nature undisturbed by humans until the stealthy moment of death.

Another foundational irony of taxidermy is that the animal's resurrection depends not only on its death, but also on the body's complete dismemberment and rearticulation. For example, in mounting a bird the animal's internal organs, brains, eyes, and so on are removed. Its legs are dismembered at the hip joint. The head is cut off. Its wings are also detached, as is the tailfeather unit. Ultimately all of these components except the soft tissues are reassembled. The wings can be shaped into an attitude of flight, the head reattached at a jaunty angle. Throughout this taxidermic process of dismemberment and reassembly, the presence of the animal's skin, and sometimes appendages such as claws, hooves, and tails, is absolutely essential. This outer covering is what meets our eye and it must never be fake. Soft tissue—eyes, nostrils, tongues—can be glass, wax, or plastic, but only the actual skin of the animal will do.[2] In the skin, in the "dermis" of taxidermy, lies its authenticating ingredient.

The notion of authenticity, of physical truthfulness, emerges too in the code of ethics of the National Taxidermists Association. Members pledge, "I will refuse to alter or falsify trophy characteristics" ("Taxidermists Code"). Taxidermist Jerry Andres reports that many sportsmen literally want him to stretch the truth: "Some guys will still come in with a bass and want you to stretch his belly so big his vent is over to one side by three inches. It's an ego trip. Or they'll want you to put a little gray fox up there snarling as if he's going to eat up the world. And everybody wants their boar's tusks pulled out another inch to make the animal look meaner" (qtd. in Curtis 94). In these junctures the criterion of realism merges with that of truth in the moral sense, so that the resurrected animal must be a faithful representation of the once-living one, not an "improved" version—bigger, meaner, more aggressive, a more challenging prey. In other words, the taxidermic animal must be an authentic representation of a particular animal, not a generic one. It must accurately render and not exaggerate the size, shape, and attitude of the specific animal when it lived. The notions of what constitutes an effective rendition of this realism have changed over time.

The development of taxidermic techniques reads as a technological history of increased "realism," and in that way joins a historical/aesthetic trajectory that moves from painting to photography to moving pictures, or from wax recording cylinders to records to CDs. Taxidermic specimens are often thought of as "stuffed" animals, and in fact taxidermy did develop from the technique of stuffing animal skins with sawdust and straw.[3] Once the skins were stuffed as full as possible, the forms were then beaten into shape with clubs. In the latter half of the nineteenth century, techniques developed rapidly, especially following the establishment of a taxidermy department in Ward's Natural Science Establishment in 1873, and the subsequent founding of the Society of American Taxidermists in 1880.[4] Stuffed animals gave way to straw models covered with clay, over which the skins were draped. Later, hollow wooden forms were used, then hollow plaster-of-Paris models, reinforced with wire. In the early part of the twentieth century, these plaster models were hailed as "the most perfect and the most scientific method up to date, [unlikely ever to] be improved on" (Shrosbree 344). During the same period this new plaster technology combined with that of early film as taxidermists turned from sketching to cameras to more accurately capture movement and muscle. In the narrative of progress, the 1970s are reported as another watershed. During this period premodeled urethane head forms were introduced, making it unnecessary to model the muscles in clay over the plaster. This streamlined the process and contributed to a growth in the number of taxidermists nationwide, extending to part-timers and hobbyists. In 1972, the National Taxidermy Association was founded to professionalize the field and to disseminate information on the latest techniques, resulting in heightened standards of realism. Commented Terry Erhlich, editor of *Taxidermy Today*, "now anybody can do passable work . . . [anybody] could mount a pheasant that looked exactly right" (qtd. in Curtis 93). Today, the latest techniques combine freeze-drying with wire supports and interior mannequins (false bodies). In taxidermy manuals, each of these technical innovations was hailed as more realistic than the preceding method, allowing for a better "fit" of skin and mannequin, or a more detailed rendering of muscle and sinew under skin.

Many taxidermists are now expanding from the static animal on a polished wood base to a narrative moment with the invocation of a habitat, even for home displays. For example, in a description that recalls museum habitat displays from early in the century, Terry Erhlich notes that "now you have a pheasant fighting with another pheasant with the hen looking on, and they are all surrounded by bushes, and grass, and gravel." In the quest for innovations in realism, animation of these mounted bodies may be coming. Says Erhlich, "I look at some of these mounts today and wonder when the thing is going to blink, or snort, or drool. Someday it probably will" (qtd. in Curtis 93).

The best advice professionals can offer the individual consumer in choosing a taxidermist is to look carefully at the work before agreeing to have anything mounted. "Do his or her mounts look alive? Do they look real?" Only an affirmative answer should lead to a purchase of services ("Picking a Taxidermist" 94). This quest for "aliveness" is taught as the guiding principle in taxidermy schools today. Some, like the Northwest Iowa School of Taxidermy, provide living animals in "live study pens," so that students can observe, photograph, and sketch deer, ducks, turkeys, and fish. In these sessions, the animal is said to "teach students proper ear positions, eye shape, and facial expression" (Northwest Iowa). Dismembering and reassembling then becomes a process of both creation and re-creation: "Skinning, tanning and mounting the skin of a mammal makes you think you're creating a live animal. Add leaves or a snow base and it is like putting the animal back in nature," asserts the advertising for the John Rinehart School of Taxidermy in Janesville, Wisconsin (*Teaching the Business*).

Ironically, it is not only humans who find the realism of taxidermy compelling in its rendering of a "real." A recent study in the *Journal of Wildlife Management* noted that live animals often mistake their stuffed counterparts for "real" species-mates. Maryland-based hunters who use taxidermic Canadian geese for decoys instead of plastic ones can catch geese more easily because the live goose will trustingly swim in closer to the decoy flock, giving the hunter a surer, closer shot. Here the resurrected dead animal attains a standard of realism convincing enough to the goose that it entices its living counterpart to the same fate (Harvey, Hindman, and Rhodes).

But if the history of taxidermy is generally told as a narrative of progress, that is, of moving from less "realistic" to more "realistic" renderings of the living through manipulation of the dead, there are two strands that contravene this narrative: fantasy taxidermy and fish taxidermy. In the 1930s and 1940s, there was a proliferation of fantasy pieces, "party-going, poker-playing rabbits, and squirrels . . . and chipmunks all dressed up in tuxedos and ball gowns." This craze has been attributed to the home-instruction manuals of J. W. Elwood's Northwest School of Taxidermy, which were published in the early 1900s and remained popular for decades. Elwood's books "would teach you how to stuff a frog and prop it up to hold a miniature beer mug" (qtd. in Curtis 64). In addition, parts of one animal could be grafted onto another, creating fantasy joke creatures like "jackalopes," which combine jack rabbit bodies with antelope antlers.

Whether created by adding props or wedding body parts across species, these variations on the "realism" theme were nonetheless dependent for their comic power on the perceived authenticity of the animal body parts and their incongruity with other body parts or fashion accessories. These displays had to look like a "real" frog drinking a beer or wearing a dress, something that a "real" frog would never do, and the discharge of laughter comes from the clash of the believable (frog body) and the unbelievable (behavior). These neo-realistic fantastical creations undercut the framing paradigm of "realism" while relying on it for their effects.

The narrative of realism also takes a surprising turn in contemporary fish taxidermy. Long regarded as one of the most difficult tasks for the taxidermist, fish mounting often accounts for half of the instruction time in a full course in taxidermy covering bird mounting, deer mounting, and medium-sized mammal mounting as well.[5] A recent shift in fish taxidermy from fish-skin-covered mounts to plastic "replica" mounts, which include no parts of the memorialized fish whatsoever, has pushed the boundaries of the profession, taking the "dermis" out of taxidermy.

In the past, fish replicas were a cheap substitute for mounted fishes, and with "a straight pose, paint on only one side, stiff fins, no teeth, and a puttied-up mouth" were easy to detect because of their insufficient attainment of current aesthetics of realism (Schultz 52). The newest versions include an impression of movement, full-body coloration, flexible fins, convincing teeth, and a detailed open mouth with tongue. (Gills remain a problem, even for the best replica makers; the sensitive moving membranes still look too rigid.)

To achieve these effects, today's molds are cast from dead fish, and each mold is used to make replica bodies, or "blanks," which are then fitted with fins. These are then painted in great detail using airbrush techniques to replicate as nearly as possible the scale and color patterning of the specific fish the taxidermy client caught. Still, even in this plastic-dominated "virtual-reality taxidermy" process, the body of the caught fish, or rather, of a fish *like* the caught fish, remains the ultimately irreplaceable origin of the "realism" of any fiberglass mount.

The utopianism of taxidermy's quest for "realistic replication" would appear to reach its current apex in cloning. But cloning reproduces living beings, and the taxidermic impulse runs precisely counter to this, in that "liveness" must remain a fiction for the uses and pleasures of such representations to be realized. The hundred-year trajectory of ever more "realistic" taxidermic representations is the history of the drive to mask the fact of death ever more completely while never completely overcoming it.

Humans after Death

The "art and science" of taxidermy, as it was referred to in the nine-teenth century, is limited to animals, however. From snakes to insects, to lob-sters, to birds, to fish, and to mammals, almost every kind of animal body has been treated in this way. But human bodies are rarely presented for public dis-play in artistic or scientific realms unless they are living, as was the case with numerous "cultural" exhibits of non-European people at nineteenth- and early twentieth-century expositions. Exceptions include some bodies in reduced forms, like mummies and human skeletons in the collections of natural history muse-ums (and ownership of these bodies is increasingly contested by indigenous peoples). So the display of dead animal bodies, culturally sanctioned through discourses of art, home decor, science, and manhood, is one of the defining lines of division between humans and animals. In this next section I want to consider the implications of this convention and some of the exceptions to it.

In general, the public display of human bodies after death in the twentieth-century U.S. is limited to the funeral parlor or occasionally to the home, when a funeral service or wake is conducted there. In Paris a century ago, however, this was not the case. Then the public morgue was open to crowds of onlookers who were ostensibly there to help identify the bodies of unknown men, women, and children displayed on slabs behind a large viewing window which could be shut off by curtains. As Vanessa Schwartz has demonstrated, such spectacles embedded these anonymous bodies in implied narratives of sensationalism, brutal acts of murder, or frightening accidents, and should be seen in relation to other sensational spectacles of the time period, like wax museums.[6] In these cases, the facticity of the dead bodies was important, but more important was the fact that, because they were anonymous, they could be displayed. Anonymity removed them from the normal conventions of death, placing them closer to the category of non-human.

This conceptual category of the non-human or the not-fully-human is what enabled the collection of tens of thousands of Native American skeletons prior to the twentieth century. Sometimes these skeletons were put on display in museums, but most of the bodies were kept in storage for use by researchers. The Native American Graves Protection and Repatriation Act, a response to indigenous activism, is now facilitating the return of these remains to their communities. These collections parallel the massive numbers of animal skeletons stored around the world, especially at university and museum laboratories. (The U.S. National Museum of Natural History, for example, holds more than half a million animal skeletons in its study collections.[7]) In addition to human skeletons, mummies (dried dead bodies with skin and organs intact) are also sometimes displayed, in museums and elsewhere, even in gift shops such as the Olde Curiosity Shop in Seattle, which displays two desiccated mummies named "Sylvester" and "Sylvia" to attract browsers (*Mummies*). Here again the distance provided by time, anonymity, and sometimes racial difference serves as the rationale for allowing such display. The conventions of treatment for these bodies are more akin to those for animals. The "thingness" of the less-than-fully-human facilitates the display.

Whereas the anonymity accorded Native American skeletons by Euro-American scientists facilitated their collection as "specimens" en masse in museums, in some cases it is not anonymity but its opposite that underwrites the display. For the famous, even a fragment of a body can take on enormous meaning, like the mummified right hand of King Stephen I of Hungary, who died in 1038 and was later canonized. The hand is on display in the Basilica of St. Stephen in Budapest. Cut off at the wrist and dark brown with age, the "holy right," as the revered object is known, is prominently displayed in a sparkling lighted glass casket ornamented with gold.

While this example is clearly tied to a Christian religious convention of reliquary, secular remains of the famous can also be displayed. Lenin's body has been preserved in a glass-covered casket since his death in 1924. Lenin's long-lived death demonstrates that techniques do exist for the maintenance of human bodies which could rival taxidermy in longevity of specimen preservation.

Lenin's body has been attended by a team of embalmers only recently released from pledges of secrecy about their techniques. Once every eighteen months, Lenin's whole body is immersed in a glass tub of chemicals containing a mixture of glycerol and potassium acetate. After soaking in this tub for two months, the body is removed, wrapped with rubber bandages to prevent leakage, reclothed, and returned to public display. The result, which I saw in Moscow in 2001, really is phenomenal. Bathed in bright light in a darkened room of his mausoleum, Lenin's pale white face and hands appear to float suspended against the dark blue of his suit in a glass casket. The clash of dark and light seems to animate the face with the potential for movement, as if at any moment he might awaken from a dreamless sleep. In the mid 1990s officials at the Institute of Biological Sciences in Moscow indicated they were hoping to market the secrets of this super-long-lasting embalming technique to rich Americans and others who might want to embalm their relatives for the long term (Tanner).

Other less well known individuals have similarly been preserved for multiple decades. Charles Henry "Speedy" Atkins was finally buried in 1994, sixty-six years after his death in Paducah, Kentucky. "Speedy" was the result of a secret embalming technique developed by the funeral home owner A. Z. Hamock. Hamock's wife Velma commented, "Speedy's never been duplicated, he's the only one that we know of. He's not stinking, nothing. The amazing thing is he really hasn't lost all of his features" ("Black Man" 56). Unknown in life, Speedy became a celebrity after his death, appearing on national TV and being featured in newspapers and national magazines. While he mostly "resided" in a closet in the funeral home, at times he was brought out and put on display for tourists, although no money was charged to see him. "I never saw a dead man bring so much happiness to people," commented Velma Hamock ("Black Man" 56).

Embalming, the removal of blood and body fluids and their replacement with a solution that binds protein and prevents bacterial growth, is usually required in the U.S. for all human bodies that are not cremated, yet few are preserved for long-term display as were "Speedy" Atkins and Lenin.[8] The purpose of embalming, according to the funeral industry, is "sanitation, preservation, and restoration to a natural appearance," all of which could also apply to taxidermy. Embalming fluids come with catchy marketing names like "Rejuvinol" and "Rejuvatone." Skin color is improved with cosmetics. At times skulls are remodeled with clay, such as Dodge's "Feature Builder," to fill in sunken cheeks and boney hollows. At these moments the process more closely parallels taxidermy, in which the skin is stretched over plastic forms, and soft tissues like noses and mouths are carefully molded (Tisdale).

A "restoration to a natural appearance" would seem important only when the body is going to be seen. But this seeing must usually be limited to a day or two's viewing at the funeral home. Clearly, although human embalming techniques exist that can provide for long-term preservation and display (as "Speedy" Atkins and Lenin prove), such displays, unlike the taxidermic displays of dead animals, are only rarely considered appropriate.

One reason for this is that the process of taxidermy is primarily a re-creative one, not a process of bodily conservation as is embalming. Replication, rather than preservation, is what matters in taxidermy. In addition, with a few exceptions like Roy Rogers's horse "Trigger" or General Sherman's horse "Winchester," who is on display in the Smithsonian's National Museum of American History, taxidermy is reserved for "wild," not domestic animals (Fleischman). It allows us to get close to animals we could not otherwise be close to—not cows, pigs, and horses, but grizzlies, moose, and wolves. With pets, however, the border between the human and the non-human is sometimes blurred.

The Special Case of Pets

With pets often regarded as "family members," their treatment after death sometimes parallels that of human bodies, with embalming and burial.[9] However, some wish to display a pet in the home, and pursue taxidermic options so the now-deceased cat can continue to "sleep" curled up in his or her favorite spot. Freeze-drying, a new form of taxidermy, is one of the most expensive options for this type of long-term preservation. Procedures can range in cost from $350 to $3,000 and can take up to nine months to complete. Few regular taxidermists will work with pets, however, despite many requests.[10] They fear that owners will only be disappointed. The live animal will obviously not inhabit the taxidermic body. While the iconicity of a mounted deer head can overcome the constant reminder of the fact that the animal died to provide the décor, with pets the situation is different. The pet's body references the pet's being, while the hunter's trophy references not only itself but the owner's feelings about himself and about hunting.

For those to whom such techniques seem ghoulish, the line between appropriate treatment of wild and domesticated dead animal bodies has been crossed. Just as displaying one's human family members in the front hall, like "Speedy," is unthinkable for most of us, so too is the display of a particular, individualized, humanized animal, a pet.

If pet taxidermy is unacceptable, another way for the body to be preserved and displayed is through pet mummification. Although practiced widely in Egypt for cats in the centuries before the birth of Christ, mummification as a burial process in the contemporary U.S. is very rare. However, catering to the growing market for pet body preservation, the staff of Summum, a non-profit religious group based in Salt Lake City, Utah, will mummify your pet. Their "certified Thanatogeneticists" will revive the "science of mummification" and wrap your pet "in fine linens bathed with fragrant herbs, oils, and resins." After two to four months, they will return your pet's body to you in a bronze "Mummiform" or statue in the shape of a dog or cat, for display in your home. (Custom designed statues looking just like your pet can be had for $14,000, but the standard fee for mummification and a generic dog or cat statue runs about $6,000.) Here the body of the dog or cat is transformed into a statue of itself, but not a mere replication since the "original" remains reside inside the statue, and the bronze becomes the new "skin" of the pet ("Mummification").

While divisions between human and animal bodies prevail in more traditional after-death practices, the Thanatogeneticists at Summum enforce no such distinctions. If you would like Summum to "lift you and set you into the hands of timelessness," you too can be mummified, so that it is possible "for you and your treasured pet to be together eternally." Prices for this are available only on request, implying that the process is less straightforward, rarer, and more variable for humans than for pets.

For those for whom the eternal promises of mummification are not enough, a new technological breakthrough holds out the hope of pet immortality. Several new companies now offer the possibility of cryogenically preserving your pet's DNA for future cloning. "Providing you and your pet the possibility to participate in the future of genetic engineering," is how Lazaron BioTechnologies of Baton Rouge, Louisiana, describes its mission, noting that although pet cloning is not yet a reality, they expect such a service to be available in the next decade (Lazaron BioTechnologies). A second business takes a more witty approach, naming itself "Genetic Savings & Clone."[11] Such potentials make the fictions of liveness more metaphysical because the fiction is that the pet itself, not just the pet's body, could be duplicated. Lazaron's promotional material offers several reasons that owners might want to clone their pets: "I could have a 'twin' of my pet when I retire in 20 years and have the time to enjoy him." "I do not know how I could ever reproduce [my horse Malibu's] unique character without cloning." That fiction represents the utopian desire of those pet owners willing to spend $500 for the "Tissue Retrieval Kit" for their veterinarian, and the $120 yearly "storage fee" for a minuscule scrap of skin. Better than freeze-drying, better than taxidermy, better than sixty-six years of embalming, cloning offers the possibility of re-created liveness—in this case the fiction is that such an animal would be the "same" animal as the pet that was cloned.[12]

Displaying the Dead in Art

As the preceding discussion makes clear, the range of morally and legally acceptable practices associated with the preservation and display of dead animal bodies is wider than that for human bodies. This is true in the art world as well as in the worlds of museum displays, interior decorating, and funereal practices. While the "art" of taxidermy has historically hovered between educational craft and aesthetic décor, it has recently been taken up by experimental artists in the visual arts. Taxidermic and preserved animal bodies have now moved from the den and the museum diorama into the art galleries of New York and London. Many of these provocative pieces, as Steve Baker has persuasively argued in his essay in this volume and elsewhere, are dependent on the literalism of the animal body as well as the de-contextualization of that body from its conventional habitat ("nature," a farm, a butcher's shop) and its recontextualization in the symbolic space of the art gallery. Hiroshohi Sugimoto, Candida Hofer, Jordan Baseman, Damien Hirst, and Annette Messager are among those artists using taxidermic animal bodies as part of contemporary installation art pieces.

Until recently, such radical recontextualization for artistic purposes was deemed immoral when it came to human bodies. The condoned public display of dead human bodies was largely limited to educational settings like natural history museums that might display mummies or skeletons, such as those found in the Hall of Osteology in the National Museum of Natural History to depict human evolution from apes.[13] In addition, collections of medical "specimens," including body parts such as heads, hands, internal organs, and so on, are used for medical education. The sight of these bodies is limited to a select audience, scientists and medical trainees, and then only under certain conditions. In earlier periods similar collections were displayed more publicly as "curiosities," especially when they represented anomalies. Today's parallel would be the pathology lab.[14] Sometimes, though, these anomalies are displayed more publicly, but still under the discourse of medicine and public education. The National Museum of Health and Medicine in Washington, D.C., is not usually on the tourist's "must see" list along with the White House and the Smithsonian museums, but it does exert its own appeal. Exhibits feature human artifacts and oddities, including an "amputated leg showing the effects of elephantitis, an esophagus with dentures still lodged midway, one thoroughly tattooed [human] skin, [and] shattered thigh bones with bullets still embedded" (Smith 7E). Although all of these contexts depend to some extent on visual spectacle and evince an aesthetics of display, rarely has the dead human body been used explicitly as a part of "art." However, a recent controversial art show in Germany walked a tightrope between the discourses of art and education and earned wild popularity.

Dr. Gunther von Hagens has developed a technique called "plastination," in which bodily fluids are removed and replaced with a molten plastic material that then hardens, so that body parts retain their color and shape. Plastination is used to create long-lasting, dry cadavers for use in medical training, but von Hagens has also used it for what he terms "anatomical artwork." In 1998–99 von Hagens organized an exhibit called "Human Body World" at the Museum of Technology and Work in Mannheim, Germany. The exhibit includes the bodies of men and women who donated them for this purpose. The bodies are posed in everyday activities like running, making music, and playing chess. The skin is stripped or flayed away, and often, as in the "Muscleman" piece, the muscles are pulled away from the bones, exposing the layers of muscles and tendons, much as did the anatomical drawings of sixteenth-century Europe. In a piece called "The Expanded Body," sections of the body are exploded outward, so that people can see what lies beneath the skull and in the thoracic cavity.

Although Catholic and Protestant church leaders have denounced the exhibit as disrespectful to the dead, popular response was overwhelming and very positive. In the first two months alone, more than two hundred thousand people attended the exhibit, some waiting up to three hours to be admitted. Some of these have signed up to donate their own bodies for similar exhibits (Andrews). Due to the demand, the museum remained open twenty-four hours a day. The church protests as well as the extraordinary audience response mark just how deep is the divide between the accepted use of dead animals for educational and artistic displays and that of dead humans exhibited under the same rubrics. The human body stands as a relic of a person, whereas the reconstruction of the animal body through taxidermy often takes on an overwhelming facticity—it becomes a specimen, standing for itself or for a category of animals like it, and not for the "being" which "inhabited" the living body.

Animatronics

The fictions of liveness that underpin taxidermy find a new outlet in animatronics. Just as taxidermy is about re-creation, not preservation, animatronics are less about mechanically replicating animals than about creating human-made versions of animals, both those that exist and those that used to or never have. In the cases where animatronics invoke animals that really exist —dogs, cats, gorillas, sea slugs—a constantly rising threshold of realism demands that animals look very "real" in order to facilitate their performance of non-realistic emotive behaviors. These articulate bodies replicate animal movement while at the same time often falsifying it: for example, providing visions of anatomically correct pigs that sing or dogs that weep.

In these instances, the realistic look of the outer layer of the body—fur, skin, whiskers, and so on—is extremely important, as in taxidermy, but it is not necessary that the outer skin be that of the animal displayed. For example, for the film *Mighty Joe Young,* about a gorilla, a team of six wigmakers spent "ten months sewing thousands of individual yak hairs onto artificial skin" to make a convincing rendition of gorilla fur (Horn). This desire is to produce a convincing replica that can then be controlled in ways that animal actors cannot. Just as the displays of taxidermy articulate a vision of science, humans, and a natural world through their precise history of staging techniques, so too does the development of animatronics exhibit a drive toward ever more accurate physical rendering to enable fantastical narrations of animal-human relations.

Animatronics have developed rapidly in the last ten years and now, in conjunction with "CG" (or computer-generated effects), have given rise to a plethora of films, TV specials, and commercials that include *Animal Farm, Babe, Jurassic Park, Terminator 2, Men in Black, Alien 3, Outbreak, Operation Dumbo Drop, Congo, The Lost World, Doctor Doolittle, Stuart Little,* and the Budweiser singing frogs. Animatronics are also popular at theme parks, originating at Disneyland in the 1960s and continuing today in ever more elaborate setups such as that at Universal Studios Escape in Orlando, Florida, which features "The Jurassic Park River Adventure." Advertisements promise that visitors to this "world of T-Rex" can "pet a living, breathing dinosaur." State-of-the-art robotics allow the creature to respond to human touch (Carstens-Faust; Universal Studios). In addition, restaurants, trade shows, shopping malls, and even celebrity homes feature animatronics as advertisement or entertainment devices.

The Development of Animatronics

It is a long way from four macaws singing calypso to the giant touch-respondent T-Rex of the Jurassic Park ride. This technological and imaginative journey began in the early 1960s with Walt Disney and then, with exponential acceleration, massive inflows of money, and technology adapted from NASA, reshaped movie-making and our fantasies of animals in the 1990s. In June 1963,

Disney opened his "Enchanted Tiki Room" in Disneyland. This generic tropical fantasyland featured singing orchids, pulsing fountains, chanting wooden "Tiki" gods, and seventy performing birds made of plastic covered with real feathers. Tape recorders hidden in the avian bodies sent electronic impulses to built-in air cylinders, creating sound and movement. The birds "sang" through vibrating reeds activated by forced air, opening their beaks, turning their heads, and flipping their tails in time to the music. Audience response to these "Audio-Animatronic" devices was immediately positive. The culminating number in the seventeen-minute show featured a slowly descending "Bird-Mobile" bearing yellow and white cockatoos singing "Let's All Sing Like the Birdies Sing." The audience went wild. This show had cost Disney approximately a million dollars and two years' time to produce (DeRoos; Anderson).

Disney saw this as a three-dimensional extension of cartoon animation, but we can also see it as harking back to the trajectory of fictional "liveness" that animates taxidermy, right down to the use of "real" feathers on fake plastic bodies—a real "dermis" over a human-made interior. What taxidermists had tried to achieve with their repertory of classical poses and their implication of narrative through the evocation of a moment frozen in time, the animatronics designers were attempting through digital controls and pneumatic valves.

Disney's audio-animatronics were not limited to animals. His first successful attempt at a human figure was a robotic Lincoln for the Hall of Presidents at Disneyland. Completed in 1964, it was much more complex than the singing birds, and relied on a newly declassified system for programming taken from submarine technology.[15] The Lincoln figure relied on analog and digital signals to control hydraulic and pneumatic valves producing movement and sound. After multiple phases of redesign involving engineers, precision machinists, sound designers, and figure makers, Lincoln could make fifteen facial expressions, lift his tongue while speaking, rise from a chair, and raise each eyebrow separately, all while delivering excerpts from his famous speeches (DeRoos; Anderson). But even though human animatronics have improved since then, they still lag behind animal developments, perhaps because we have a higher threshold of realism for replication of humans than of animals.

These early robotic figures would seem clunky and unconvincing to us today, with their limited range of movement and lack of modulation in acceleration and deceleration. In early experiments, shooting films with robotic figures proved unwieldy and slowed the pace of filmmaking. For animatronics to move from Disneyland into the movies, new combinations of expertise were needed. Special effects technicians who excelled at building prostheses (fake body parts) became involved. Mechanical engineering melded with the techniques of puppeteering, as workers in both Britain and the U.S. experimented with new ways to create animals that moved. In the late 1970s and early 1980s, makers of radio-controlled models, computer programmers, and makeup artists were pulled into these circles of free-lancers working in experimental labs (Maley).

Getting the creature's physical representation accurate enough to convince the viewer's eye in closeup is one of the key challenges of producing animatronic figures for films. Skin and eyes are especially telling, but musculature and mass are also important. Movement, articulated by muscles and displacing mass, always poses a difficult problem.

Making Animals Move

Human and animal movement is extraordinarily complex. Robotics designers estimate that human movement has more than two hundred degrees of freedom, or possibilities for variance, while robotic structures often deal with only six. All robotics, whether generated as computer schema or mechanically built, must offer a schematized version of human or animal movement.[16] The question becomes which characteristics of movement must be reproduced for the end result to be effective. Even getting robots to walk on an uneven surface with a human-like gait is an enormous intellectual and programming challenge. The dynamics of human and animal movement, that is, the subtle accelerations and decelerations that yield a smooth reach for a cup of tea, or a graceful swinging from branch to branch, are very hard to duplicate.

There are two main schools of approach. One aims to analyze the movement into accurate algorithms in order to reproduce it, and the other aims to reproduce the perception of the movement. For example, devising a moving body of water could be done by studying fluid dynamics or by creating a mathematical model for a dappled surface which appears to reflect light in the same way as a moving current. In both cases, the complexity of the "natural world" remains both ultimately unobtainable and the model against which the perception of the end product is judged.

The introduction of servomotors into animatronics technology in the 1980s resulted in a new intimacy between human bodies and animal bodies, and made movement more subtle (Maley). These devices, often called by their trademarked name "Waldo," are "ergonomic-gonio-kineti-telemetric input devices" used to control animatronics. Worn on the puppeteer's body, these contraptions of metal cuffs, straps, caps, and sensors directly link the movement of the puppeteer to the movements of the artificial creature through electrical and mechanical transmission and translation. Waldos are engineered to fit a performer's body or head comfortably and to allow a wide range of movement. The angle, speed, and range of this human movement are electronically measured, and this data is telemetrically transferred to the three-dimensional animatronic creature (or to its two-dimensional computer-generated clone, an "electronic puppet"). The "Dual Arm Waldo," for example, straps onto each arm of the puppeteer, producing a cyborg-esque mix of flesh, metal, and electronics that can make an animal's limbs move. The "SimGraphics Facial Waldo" looks somewhat like a virtual-reality helmet, with sensors detecting movement at multiple points along the brows, cheeks, lips, and jaw. The performer's face controls the creature's expressions ("Waldo").

Whereas earlier puppeteering efforts in movies had required the coordination of teams of puppeteers, each juggling rods and levers or twirling joysticks to move a part of the creature's body, now just one person can coordinate the movements of multiple body parts. The interface between human movement and human imitation of animal movement is now closer. For instance, facial movements that were previously controlled by hand can now be transferred directly from the human performer's face to the creature's face via the "Facial Waldo." This makes the transformation of animal faces into the expressive coding of humans more seamless—a literal anthropomorphizing. It is important to remember that this anthropomorphism is simultaneously an animalomorphism—the human actor's body adjusts its movements to "become" animal-like as well. A mutual morphing occurs. These new techniques underlay the success of the film *Animal Farm*.

Animal Farm combined animatronics technology with live-action animal actors and computer-generated effects. In the film, it was essential that the animal actors, who dominate the film, speak in a convincing way. Live animal actors played the leads, and animatronic versions of their specific bodies were then created based on careful measurements and modeling. In effect, a mechanical clone was produced, but one which could be operated by puppeteers using motors and joysticks. Large-scale animal models allowed the puppeteers to get right inside them, literally inhabiting and animating the animal body. Humans also inhabit the animal bodies in the sense that their voices (the human voices actors provide for the speaking roles of animals) are made to fit (literally) into the animals' mouths. Actors watch the already animated mouths of the animatronic animals and time their lines and the movement of their own mouths to fit with the motions already on the screen.

When smaller models are used, they are run by servomotors from off screen. In these cases, a puppeteer's supple hand fitted with a special electronic glove controls eye movements, lip movements, ear cocks, and nose wrinkling. Close-ups of these faces allow viewers to concentrate on the emotive facial expressions of the animals, while wide-angle shots using live animals, or even computer-generated versions of the animals, are used for action scenes. The aesthetic goal is to have the intercutting of live, animatronic, and CG animals be seamless, so that none of the shots appears more "real" than another, within the already fictional framework of a talking-animal show.

Ultimately, animatronics faces the same limitation of motional accuracy that underlies taxidermy. Dead animals cannot move, but elaborate taxonomies of postures have developed to imply movement for various species, often representing a moment in time and hence a suspended narrative. Animatronics takes this one step further by supplying actual motion which represents "accurate," "realistic" movement but never completely duplicates it. In some cases the motion is purely imaginary—the animators from Industrial Light and Magic created the movement of the brachiosaurus in the movie *Jurassic Park* by spending hours at wildlife parks watching animals and touching their skin. The result is a "dinosaur that strode like a giraffe but weighed as much as an elephant, a musculo-skeletal impossibility" (Westrup 10). In these cases "realistic" does not mean anatomically and kinesthetically accurate (as if such knowledge about the

Displaying Death, Animating Life 173

brachiosaurus were even possible), but rather anatomically plausible. These "re-creations" are "re-inventions."

The production of the film *Mighty Joe Young* provides a strikingly complex example of this melding of creation and re-creation.[17] It involves a series of sub-stitutions anchored in the bodies of living animals, but transmuted through the body of a human actor into an animatronic clone of that human acting as an animal, into a computer-generated graphic clone of the animatronic version of the human imitation of the living animal—in this case the gorilla, Mighty Joe.

For most of the film Joe was played as a mature gorilla, by "seasoned ape-suit performer" John Alexander. Known as the best in the business, Alexander is re-garded as an expert on gorilla behavior. "I tried to get a picture in my mind of how Joe would move and react. . . . I found Joe through his physicality, and transferred that to my performance," explained Alexander (qtd. in Essman 92). Ultimately, it was performer John Alexander who set the limits for Joe's appear-ance. "How much he could distort his body to look like a gorilla would deter-mine what Joe would look like," stated gorilla suit designer Joe Baker (qtd. in Essman 79). The animal body, in other words, is literally created on a human form which becomes like the interior mold of a taxidermic specimen. Using special hot-melt adhesive, a gorilla body was molded over John Alexander's body posed in a gorilla stance.

For scenes where the character's actions exceeded what actor Alexander could do—for instance, smashing a Mercedes or climbing on a Ferris wheel—a com-puter-generated version of Joe had to be created. This was modeled not on a real ape but on Alexander's body moving like an ape. A similar challenge awaited in the design of the creature's CG facial expressions, which were to be more ex-treme than those in the animatronic version. "Here the challenge was not to recreate the facial characteristics of a real gorilla," remarked visual effects spe-cialist Dan DeLeeuw, "but to replicate the nature of the *animatronic* ape as closely as possible" (qtd. in Essman 88, emphasis added). For those scenes that would be shot with Alexander against a blue screen and then "matted in" elec-tronically, puppeteers guided Joe's face in a symphony of densely compacted moves. A mechanical engineer operated the nose and upper lip with his right hand and the lower jaw and lip with his left. Each hand controlled eight tracks of detailed electronic information sent through a transmitter. Three other pup-peteers coordinated movements of the tongue, cheeks, eyes, and brows with standard radio-control units. This complicated ballet of movement substituted human movement for the animal movement it supposedly replicated and en-hanced. Unlike the taxidermic evisceration of the brain, heart, bones, and guts of the animal, the puppeteers' version cognitively dissected each muscle group of the face, and then articulated them in space and time to render the fiction of emotion through movement.

In this shell game of origin and copy, or "real" and "replication," the starting point is an unseen gorilla in a forest somewhere, but in the process, the human actor as a gorilla becomes the physical and kinesthetic model for the animatronic version, which in turn becomes the model for the computer-generated version. In this conceptual and material process, the origin (some gorilla somewhere) is both material and imaginary, since the actor's embodiment of "gorilla-ness" is dependent on a memory of a perception, and an extrapolation from that memory into predicted behavior (what that gorilla might do in this situation). The final product, the two-dimensional visual rendering of a gorilla, takes the notion of bodily "truth" (what a gorilla looks like, how it moves, how it reacts) and extrapolates from it in the service of fiction for emotive effect. Taxidermy does the same, fixing the twist of a head into an expressive "typical" pose, activating for us some narrative of a life usually beyond our ken, and anchoring it in the physicality of the now reanimated remains.

All of these techniques involve displaying dead bodies of real animals or creating fake bodies of animals meant to look "real" or at least real enough. These acts of representation engage various (and changing) technologies of realism to enable elaborate narrations of social relations, whether we are considering the staging of natural history dioramas or the freeze-drying of pets or the design of a gorilla's facial movements. They evidence elaborate passions for control over animals as both our closest interlocutors and our always non-human "other." The ultimate emphasis in these categories of representation is on an outer physical rendition that is persuasive of an interiority that is not animal but is rendered by and for humans.

Despite the obvious differences between animatronics and taxidermy, both involve a complex struggle to re-create animals. Skeletons are built or rebuilt from scratch, movement is intimated or added. An intense intimacy between animal bodies and human ones results. In taxidermy this involves human manipulations of dead animal skins. In animatronics it involves human bodies moving inside outer mechanical shells, inhabiting an animal body. This intimacy is always simultaneously marked by the distance and distinction between animals and humans. It is the rendering of that distance and the continuing desire to overcome it that make such fictions of liveness both possible and desirable, even necessary.

Notes

My thanks to graduate students Christine Bruger and Bill Bryant, who assisted with this research with energy and excitement. My conversations with them stirred my thinking, and many of the leads they dug up helped make this article possible. Thanks also to museum design specialist and artist Bruce Sherting for stimulating discussions and for leading me to Gunter von Hagens. Virginia Dominguez provided stimulation and support, as always, and is a continuing interlocutor in my passions about animals. Thanks to Nigel Rothfels and Andrew Isenberg for putting together the very exciting conference "Representing Animals" at the University of Wisconsin–Milwaukee that encouraged this work. Conversations with the participants at that conference also helped shape my thinking, and I thank those who shared thoughts with me, including Molly Mullin.

1. In using the terms "animal" and "human" instead of "non-human animals" and "human animals" throughout this piece, I am stressing the epistemological division that the non-overlapping terms imply.

2. In some of today's mounts, even the skull is omitted. In a modern deer head mount, for example, only the skin and antlers are used—eyelids are sculpted from clay, the nose and mouth from wax; eyes are made of glass; and the mannequin or form over which the skin is draped is made of polyurethane foam ("General Information").

3. We retain the term "stuffed animals" for children's toys. In a disconcerting irony that echoes the foundational irony of taxidermy, the Smithsonian Institution recently sold stuffed animals made by Steiff (all acrylic fur) in its limited-edition Animal Conservation Collection. Monies from the sale of these very expensive toys ($225 for a leopard, $300 for a sloth), which reference taxidermy in miniature, went to support the National Zoo's conservation studies.

4. Graduates of Ward's included William T. Hornaday and Carl Akeley, central figures in the development of the natural history museum. The history of taxidermy is tied up with that of natural history museums, but is not limited to them. Entertainment displays at expositions provided early outlets for pleasurable viewing based more on spectacle and less on sanctioned "scientific" looking. Taxidermy became more popular as museums expanded from scientific collections used by researchers to institutions for public education. Increased drama in taxidermy stagings facilitated this transition. In addition, taxidermy has a long tradition as an amateur hobby, especially for boys. During the Victorian era, natural history objects, including stuffed birds and small mammals, were a popular part of home décor.

5. This information is from training schedules listed in the promotional materials (for the year 2000) of the John Rinehart Taxidermy School in Janesville, Wisconsin, and the Colorado Institute of Taxidermy Training, Inc., in Canon City, Colorado, both of which devote the largest portion of their training days to fish taxidermy. Since general taxidermists will spend most of their time mounting deer heads, the length of instruction is not tied to the expected volume of business, but rather to perceived difficulties in obtaining a sufficiently "realistic" rendition of fish.

6. See Schwartz, especially chapter 2, "Public Visits to the Morgue: Flânerie in the Service of the State." I thank Connie Berman for bringing this book to my attention.

7. See Hafner et al. I thank Erika Hill for this reference.

8. Some religions, like Orthodox Judaism, prohibit embalming.

9. In a separate essay in progress, "On the Margins of Death: Mourning and Funeral Displays in Pet Cemeteries," I develop issues related to pet burial and pet cemeteries, but those issues exceed the limits of this essay, which focuses on visual display of the bodies themselves. While pet cemeteries rarely embalm animals, some will, thus ensuring a longer-lasting bodily presence on this earth. In 1995 Peter Drown, then head of the International Association of

Pet Cemeteries and a former human embalmer, would embalm a pet cat for under $100 (Waytink).

10. Animal Art Full Service Taxidermy Studio in Cordova, Tennessee, is an exception. They will make taxidermic mounts of domestic pets for a minimum charge of $550 plus $15 for every pound over the first twenty pounds (Animal Art).

11. Some businesses promote their services not just to pet owners but to animal breeders as well. PerPETuate, Inc., of Farmington, Connecticut, offers a "head start on cloning your prized animal," whether that be a guide dog or a top-producing dairy cow (PerPETuate).

12. No such cloning services yet exist for humans, but recent advances in tissue preservation have opened up new possibilities. Cryogenic preservation of humans is available for a (hefty) fee. The Alcor Life Extension Foundation in Scottsdale, Arizona, will place a human in cryonic suspension—quick-frozen in liquid nitrogen—in the hopes of later being "reanimated." The cost is $120,000 for a full body, and only $60,000 if just the head is stored. Perhaps the assumption underlying the bargain of the "head only" is that by the time you are unfrozen, scientists will have figured out how to fabricate the rest of the body and you can be reassembled with your head and a bionic body (*Mummies*).

13. The display of dead bodies also has a long history as a way for one group to terrorize another. In these cases the dead serve both as evidence of the infliction of power and as warnings to those who see them. Lynchings in the United States and the killing fields of Kampuchea are among the horrific examples. While the displays of dead humans I discuss in this paper also exhibit the operation of power, in most cases it is not within an explicit discourse of terror.

14. Medical education utilizes human cadavers and body parts preserved in fluid. Generally these are not on public display outside of the classroom, though. The aesthetics of dead body display and the discourse of anatomy that surrounds it in medical training deserve further investigation. A comparison with veterinary training would be revealing. Sometimes human "medical curiosities" were displayed after death outside the medical teaching realm as well, for "entertainment" purposes in circus sideshows or other displays. The dead body of "ape woman" or "bearded lady" Julia Pastrana, which was displayed in fairs and carnivals during the latter half of the nineteenth century, is just one such example. I thank Nigel Rothfels for bringing her life history to my attention (see also Bondeson).

15. Earlier efforts had already adopted the use of the inertial reference integrating gyro, recently declassified after its development by NASA, linking the national fantasy of moon travel with the fantasy of creature re-creation.

16. My thanks to robotics specialist Joe Kearney for describing certain aspects of his work to me and for loaning materials and suggesting examples like the water dynamics that helped me frame this discussion (see Badler, Barsky, and Zeltzer).

17. This 1998 film is a remake of the 1949 *Mighty Joe Young*, in which a gentle gorilla ends up on a rampage in Hollywood. The special effects in that film were produced by painstaking frame-by-frame manipulations of a puppet of

a gorilla moving across miniature cityscape sets. Fifty years later, the remake was built around a man in a gorilla suit, animatronic gorillas, and an assortment of dazzling computer and matting effects.

Bibliography

Anderson, Paul F. "Audio-Animatronics." *Persistence of Vision*. Accessed 28 Jan. 2002 <http://www.disneypov.com/issue06-7/aa.html>.

Andrews, Edmund L. "Anatomy on Display, and It's All Too Human." *New York Times* 7 Jan. 1998, late ed.: A1.

Animal Art Taxidermy Studio. Accessed 29 Jan. 2002 <http://www.animalartonline.com/prices.htm>.

Badler, Norman, Brian Barsky, and David Zeltzer. *Making Them Move: Mechanics, Control, and Animation of Articulated Figures*. San Mateo, Calif.: Morgan Kaufmann, 1991.

Baker, Steve. *The Postmodern Animal*. London: Reaktion, 2000.

"Black Man Who Died 66 Years Ago Is Finally Buried." *Jet* 29 (Aug. 1994): 56–59.

Bondeson, Jan. *A Cabinet of Medical Curiosities*. Ithaca: Cornell University Press, 1997.

Carstens-Faust, Jill. "Off to the Islands." *Home and Away* 20.5 (Sept.–Oct. 1999): 44.

Curtis, Sam. "Whatever Happened to Taxidermy?" *Field and Stream* Aug. 1994: 62+.

DeRoos, Robert. "The Magic World of Walt Disney." *National Geographic* 124.2 (Aug. 1963): 159–207.

Essman, Scott. "A Gorilla Named Joe." *Cinefex* 76 (Jan. 1999): 76–104.

Fleischman, John. "The Object at Hand." *Smithsonian Magazine* 27.8 (Nov. 1996): 28–30.

"General Information." *Taxidermy.Net*. Accessed 28 Jan. 2002 <http://www.taxidermy.net/information/whatis.html>.

Hafner, Mark, et al. *Mammal Collections in the Western Hemisphere: A Survey and Directory of Existing Collections*. N.p.: American Society of Mammologists, May 1997.

Harvey, William F., IV, Larry J. Hindman, and Walter E. Rhodes. "Vulnerability of Canada Geese to Taxidermy-Mounted Decoys." *Journal of Wildlife Management* 59.3 (1995): 474–77.

Horn, John. "Gorilla Warfare." *Premiere* Jan. 1999: 51–53.

Laughing Elk Studios. Accessed 31 Jan. 2001 <http://www.animalmounts.com>.

Lazaron BioTechnologies. Promotional kit. Baton Rouge: Lazaron BioTechnologies, 2000.

Maley, Nick. "Movie Animatronics vs. Disney's Audio Animatronics." Accessed 28 Jan. 2002 <http://www.cinesecrets.com/pmDisneyAnimatronics.html>.

McKenzie Taxidermy Supply. Accessed 28 Jan. 2002 <http://www.mckenzi-esp.com/8600_series.asp>.

Mummies: Frozen in Time. Pangolin Pictures. Copyright Discovery Communications. The Learning Channel, 1999.

"Mummification for Pets and Animals." *Summum*. Accessed 29 Jan. 2002. <http://www.summum.org/mummification/pets/>.

Northwest Iowa School of Taxidermy. Accessed 28 Jan. 2002 <http://www.matuskataxidermy.com/Pages/study_pens.htm>.

PerPETuate, Inc. Accessed 28 Jan. 2002 <http://www.perpetuate.net>.

"Picking a Taxidermist." Sidebar to Curtis, 94.

Schultz, Ken. "Virtual Reality Taxidermy." *Field and Stream* Aug. 1995: 52–55.

Schwartz, Vanessa R. *Spectacular Realities: Early Mass Culture in Fin-de-Siècle Paris.* Berkeley and Los Angeles: University of California Press, 1998.

Shrosbree, George. "The Scientific Development of Taxidermy and Its Effect upon Museums." *Wisconsin Academy of Sciences, Arts, and Letters Transactions* 16:1 (1908–1909): 343–46.

Smith, Amanda. "Museum Takes Inside Look at the Human Body." *Des Moines Sunday Register* 30 July 2000: 7E+.

Tanner, Adam. "Lucky Stiff." *National Review* 7 (Nov. 1994): 32–34.

"Taxidermists Code of Ethics as Adopted by the National Taxidermists Association." *Taxidermy.net.* Accessed 28 Jan. 2002 <http://www.taxidermy.net/nta/code.html>.

Teaching the Business of Taxidermy. Brochure. Janesville, Wisc.: John Rinehart Taxidermy School, 2000.

Tisdale, Sallie. "The Sacred and the Dead: Autopsies, Embalming, and the Spirit." *CoEvolution Quarterly* 41 (spring 1984): 4–12.

Universal Studios Islands of Adventure. Brochure. Summer 1999.

"The Waldo." *The Character Shop.* Accessed 28 Jan. 2002 <http://www.character-shop.com/waldo.html>.

Waytink, Judy. "Gone, but Not Forgotten: Laying Deceased Pets to Rest." *Cats* Sept. 1995: 36–39.

Weiss, Rick. "Copycat (and Dog) Business Opens." *Iowa City Press Citizen* 16 Feb. 2000: A1.

Westrup, Hugh. "Electronic Dinosaurs Stalk Screen." *Current Science* 79.2 (2000): 8–11.

10 Bitches from Brazil:
Cloning and Owning Dogs
through the Missyplicity Project

Susan McHugh

I'm fascinated by the Genetic Savings and Clone project because it emphasizes the idea that DNA is the "bottom life" for the way animals and humans behave. But most people are smart enough to know that, as powerful as genes are, they don't explain everything. People and animals change over time. Historical circumstance[s] do too. I think those who most object to the idea of cloning are people who for whatever reason can't appreciate this.

—Stephanie Turner

In 1997, the Roslin Institute announced its successful cloning of Dolly, a Finn Dorset ewe, who is the most notorious but not the first mammal to be cloned. Still, Dolly's creation, which involved the transfer of DNA from the cell of an adult sheep into an evacuated egg cell, signals revolutionary developments in cloning: demonstrating a way to replicate adults who otherwise have no reproductive capacity, Dolly's production breathes new life into the genetic legacy of even sterile mammals just as her worldwide reception secures a newly public life for the clone in popular media. Indeed these enormous technological changes may well be enhanced by the popular presentation of this new method of cloning, for it presages major social changes in perceptions of and uses for animal clones. A few months after Dolly was introduced to the world, an anonymous married couple donated over two million dollars to found the Missyplicity Project—now a nearly four-million-dollar joint venture between the Texas Agricultural Experiment Station at Texas A&M University (TAMU) and the San Francisco–based Bio Arts and Research Corporation (BARC)—which is designed to accelerate the application of this method in the production of companion animal clones.[1] The singular name of the Missyplicity Project reflects its equally singular primary goal, which is to clone the "perfect" pet from the donors' spayed, mixed-breed bitch named Missy.

Publicly announcing its aim of turning Missy (the well-loved, aged former stray) into Missyplicity (a pack of clones genetically identical to each other and to Missy), the Project at its inception diverges sharply in substance and presentation from all other cloning projects, including those of the Roslin Institute. Instead of offering clones of exclusively "breed animals" for brief public scrutiny after their birth, the Missyplicity Project heralds its scientifically unique focus on cloning a random-bred animal through a comparatively open and ongoing narration of its progress, from its very incorporation as a research project through its as yet incomplete goal of cloning Missy. That is, the focus on an animal of uncertain origins together with the consistently public presentation of the Missyplicity Project make the cloning of Missy a culturally singular event.

By attempting to clone an animal who, from the standpoint of the burgeoning "technoscientific" animal industries, represents a non-human-regulated and therefore worthless genetic record, the Missyplicity Project, in its object-choice, moves against the tide of economically driven science in the Genetic Age.[2] But its aggressive promotion of this unique focus indicates how the Project is very much of this age. While Missy's clones will not be marketed but given strictly to predetermined owners (including Missy's own) and all additional dogs involved in the Project will be given away (yet another major difference from all comparable projects), the techniques developed through this process, involving the biologically unique and "seriously underfunded" scientific research area of canine reproductive physiology, are patentable and projected to have immediate applications in such diverse areas as service dog breeding, endangered canid preservation, and canine population control ("Press Releases"). Moreover, the Missyplicity Project anticipates the immediate development of a market in pet cloning through the birth of Missy's clones: the Project, through its offshoot, Genetic Savings & Clone (GSC), a privately owned cellular DNA storage facility and cloning service that has the same directors as Missyplicity, has staked its claim in the next stage of what Susan Wright terms "the cloning gold rush" (304). The public presentation of Missy as both beloved companion animal and clone mother conjoins these twin projects, positioning Missyplicity as a curious point of reconciliation between the work of science and the love of dogs.

Missy's dual status as a companion animal and a cloning donor proves a key component in the Missyplicity Project's extension of its use of animal representation to enable Missy to continue Dolly's work of popularizing cloning science. Specifically, the Project uses stories and images of this particular dog to turn the potentially alienating science fiction of cloning—what I term, after Ira Levin's 1976 cloning novel, "Bitches from Brazil"—into the future of dog owning. The stories and images of Missy that structure the Project's official website, *Missyplicity.com*, betray a potent collusion of scientific and pet-owning aesthetics of control, a collapse of scientific frontiers (non–breed animal genomes) into marketing niches (the newfound property of pet owners). Particularly by framing images of the lone bitch with narratives characterized by "gooey sentimentality" (Sirius), the Project's website records the replication of Missy in ways that ensure the widespread social acceptance not simply of Missy's clones but also of companion animal cloning services like GSC.

Continuously recording the progress toward cloning Missy, the Project's website repeatedly pairs images that play up the campy uncanniness of clones with stories geared to cultivate the special canine-human bond that Marjorie Garber terms "dog love." Simultaneously the Project's official story and its sounding board, the website weaves together what it terms "bi-directional communication" or two-way discussions of Missyplicity ("Welcome"). That is, it incorporates negative as well as positive feedback from readers alongside responses and progress reports from the Project's manager, all augmenting the site's collection of stories ("MissyTales") and images ("MissyMedia") of Missy the dog. Missy's necessarily mediated presence fortifies this "bi-directional" narrative strategy—in particular, the positioning of GSC as a response to warm feedback to the Missyplicity Project from pet owners, inspired by Missy to consider cloning their own companion animals—indicating how the Project's evolution into an extension (rather than critique) of the biotechnological commodification of non-human animals involves the representation of this special animal. The Missyplicity Project's website assumes that cloning can succeed as a commercial service only if it is broadly perceived as enhancing human cultures. Drawing strength from opposition to both clone and non-human animal cultures, the Project frames Missy squarely within the context of Missyplicity to position dog cloning subtly but surely as the next stage in the history of human-manipulated dog breeding.

Conceived as an integral part of the Missyplicity Project, the site's narratives of Missy, particularly the MissyTales, situate dog cloning as a method of supplementing (as opposed to creating) dog love. Under a framed portrait of Missy, against a background or wallpaper of multiple images of her, as well as diagonally below an animated image of her in which her cartoon tail endlessly wags, the "Welcome" page (fig. 10.1) of the Missyplicity Project proclaims, "Missy is a beloved pet, getting on in years, whose wealthy owners wish to reproduce her—or at least create a genetic duplicate (which we all know is not the same thing)." It thus plunges straight into the Project's central paradox, namely that Missy's cloning comes at the expense of that which motivates it, that is, Missy's unique and inscrutable genetic status as a mongrel. "Most people who meet Missy feel she is a special dog," the "Welcome" page continues, and the primary task of the Missyplicity Project is to empirically validate that pattern of sentiment.

Because the chosen method of validation involves multiplying her exactly, making her the first of a series of Missys, her "specialness" motivates a research project that will pin down and thereby establish her (for now) incomprehensible breed status. Indeed, in the website's terms, cloning promises to correct Missy's non-breed status, the otherwise irrecoverable genetic hallmark of bad (or canine self-selected) breeding in dogs. To tackle this formidable task, the site's narratives rely on a peculiar fusion of dog love, breed narratives, and visual imaging to insist that humans and dogs, Missy and her owners above all, can recover Missy's breed status, thereby retroactively justifying the mongrel's genetic fitness for cloning.

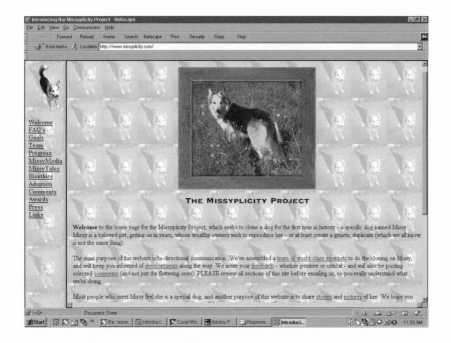

10.1. Introducing the Missyplicity Project.

Thus, as a recovery project, Missyplicity substitutes the fixity of the cloned genome (Missy) for the mutability of the mutt's breed-status (the canine social group through which she was produced). The site begins to tell the story of the transformation of Missy into Missyplicity as the necessary development of a colony of clones from this singular canine by emphasizing the unique scientific value of this cloning project in its focus on a mongrel bitch. In pointed contrast to the exclusively "breed" livestock animals already cloned, Missy is an animal of uncertain parentage. While on the one hand the Project positions Missy's mongrel status as its central value—"Missyplicity is largely a celebration of mutts" ("Frequently Asked Questions")—it also sets clear limits to this celebration. For cloning Missy holds out the bifold promise of replicating the bitch for her human owners and, in the process, reinventing this sterilized mongrel as a clone mother of a new breed, one that remains to be constituted by her clones. Given these conditions, the site has to resolve the problem of Missy's uncertain breed-status, and it does so with the term "a breed apart," a phrase that simultaneously erases Missy's mongrel status, anticipates the need for a new conceptual category to account for Missy's clones, and validates human perceptions of

Missy's own special status among dogs. Even before her actual cloning, the site thus transforms this bitch into both *über*-mutt and the clone mother of the end of all mutts.

The multi-faceted contradictions involved in articulating Missy's genetic record haunt the "MissyTales," which account for Missy's mongrel history while at the same time casting that history as the foundation of a new breed. Unrecoverable origins, arguably the defining trait of the mongrel, trouble these tales, written for the website by Missy's "'human mother' (who prefers this term to 'owner')." Significantly, Missy's tales are filtered through this "motherly" (and explicitly not fatherly) human presence, for these stories draw heavily on the conventions of maternal melodrama to reconstruct Missy's origin not at the site of her canine parents' sexual encounter but at the first encounter between woman and dog. Mary Ann Doane, in her account of the rise of maternal melodrama as a popular cinematic genre, argues that its defining tropes, namely "proximity rather than distance, passivity, overinvolvement and overidentification," contain "the woman's assumption of the position of 'subject' of the gaze" (2). Distinguished up front as the bitch's "human mother" instead of "owner" or "human father," the author of these narratives circumscribes her own subjectivity as she falls back on clichés of maternal power overwhelmed by emotion regarding Missy. In Doane's terms, Missy's "human mother" gains a certain "psychical appeal" as she writes about what she sees in Missy to account for the desire to clone her, but she gains it at the expense of "empowerment" (9), in the broadest sense.

For it is not simply the human mother's but more importantly the bitch's authority that the tropes work together here to evacuate. The first of the MissyTales, "How Missy Found a Home," moves Missy from her murky animal-shelter past into the human mother's home in a narrative power play that conflates Missy's personal and genetic histories as it dismisses them. This and the other MissyTales contrast the instantly obedient Missy with her immediate non-human predecessor within the family, a "disobedient" coydog (or dog-coyote mix). Thrown in sharp relief against the coydog, who bites, repeatedly frees herself from both collars and pens, and hunts to kill, Missy appears special from the start because she instantly exhibits human-compatible characteristics—fetching balls, barking on command, even smelling good—which inspire her human mother, as she puts it, to fall "into deep and permanent love with Missy." Within the melodramatic structures of this mother-bitch narrative, it was dog love at first sight for the long-suffering narrator, who casts her power of choice in the matter of selecting a new dog as overcome by Missy's charms. The story frames cross-species identification in intrahuman familial terms to position Missy as the adopted child of her human mother, who finds the bitch not only trained but "oh so beautiful" from the start.

The human mother's development of the dog in terms of progeny compares Missy favorably with the coydog, in relation to whom Missy appears to be a pure and superior dog. Although the stories, like Missy, remain locked in the maternal gaze, these hints of the coydog's story contextualize Missy's introduction into this home amid the failure of dog love to assimilate her non-human predecessor into the human mother's culture. Shadowing this collapse of distance between human mother and dog daughter is the question of how the coydog's exhibition of "feral" behavior (like biting, roaming, and killing) indicates that, in the terms of Missyplicity, bad dogs primarily identify with canines and good dogs like Missy invite human identification. Through a peculiarly mixed account of her biological and behavioral characteristics, "How Missy Found a Home" implicitly rejects the possibility that Missy's unrecoverable preshelter life involved active enculturation in other human homes in favor of the idea that submissive Missy was born well-behaved as well as beautiful. While humans will engineer her clones to be this way, Missy herself, the human mother asserts, was found supremely biologically destined to enjoy dog love.

Like dog-breeding stories generally, this tale compromises its own invocations of dog love to justify human intervention in canine reproduction. Garber's extended study on the subject suggests that dog love or "caninophilia" is "an erotics of dominance" that circularly functions as the means to the end of replicating dog love (125). The dog owner, finding herself in thrall to her beloved pooch, exerts her control over the animal's reproductive (and in this case replicative) capacity to produce another who will similarly captivate the owner. In pointed contrast to dog trainer and animal-behavior philosopher Vicki Hearne, who contends that the ideal of obedience to human "corrections" structures all productive human relationships with dogs (69), Garber's theory of dog love accounts for Missy's human mother's need to position the dog as superhuman in order to tell the story of her cloning. Dissimulating who serves as the ultimate judge of canine behavior, the human in dog love portrays herself falling under the spell of a transcendent dog, in this case Missy.

In this way, dog love allows a reconfiguration of social power in the tale of Missyplicity that contradicts the gross imbalance of power conventionally assumed between humans and animals in the labs of cloning experiments. And it also provides a way to reconcile the Missyplicity Project with the history of human-controlled dog breeding. Seeing Missy as the opposite of the presumably undeserving coydog, whose canine behavior oddly confirms the visual evidence of cross-species breeding, involves looking at her through the human mother's melodramatically tinted lenses. Only after the possibility of cloning arises does mongrel Missy in body and mind, in her "good" smell and her retrieving behavior alike, belie the logic of breed with her pure dog status, at least for the human mother who wants more Missys.

The MissyTale devoted to developing this cloning desire, titled "A Breed Apart," works even harder to shift responsibility for the cloning of this bitch from her owners to random human observers, who "[e]very day" marvel at Missy's beauty. The Project in this respect serves not the whim of the human mother so much as the proprietary interests of Missy's human acquaintances, who seem "ready to go immediately to get one just like her." In this story, her human mother suggests that, by cloning Missy, "her humans" (again, a term used in lieu of "owners") are not creating but merely confirming a popular perception of Missy as a "breed" dog. In short, they are only giving in to a popular demand by funding the development of clonal answers to the question that "everyone" asks: "What breed is she?" Without clones to justify the invention of this "breed apart," the question of what breeds may have been involved in the production of Missy becomes more difficult, if not impossible, to answer. Claiming that "Missy is the dog equivalent of a Rorschach test," in that "[p]eople see the breed or blend they want to see," the human mother admits "nothing conclusive to go on visually." Instead the story prefigures Missyplicity by continuing to blur together Missy's behavior and genetics or, more simplistically, her nurture and her nature, "narrowing the possibilities" by accounting for all of her desirable qualities in terms of a single, self-original breed.

The lone mutt becomes a clone mother, then, on the condition that she first becomes visible as a "breed" dog. As this tale singles out Missy as currently the sole member of "a breed apart," her mongrel status, the singular condition on which she is cloned, mutates into the genetic singularity that serves as the foundation of a new canine breed. This configuration posits an inevitable (if not quite natural) selection of Missy only to conceal the human hand that rocks the cradle of the Missyplicity Project. Whereas Missy's canine family conditioned her humble birth, which led quickly to her own sterilization, her "human family" determines her notorious genetic future, which will be conditioned by her replication as the progenitor of the one, true breed apart. And, with this incorporation into the ideology of breed, Missy's replication itself comes to replicate rather than to challenge the erasure of canine culture in human-mediated dog breeding.

However passively and passionately narrated, love for Missy traces an imbricate pattern of cloning and owning that collapses some teleological structures of breed and species only to reinforce others. Dog love leads to Missy's cloning through a certain language of control that inscribes an "erotics of dominance" that augments the human's sense of self at the expense of the dog's identity (Garber 125). Suggesting how dog cloning exacerbates this imbalance of power, Project coordinator Lou Hawthorne most bluntly clarifies the anthropocentric terms of this distinction in reply to a message from one of the site's readers:

> Everyone has agreed that cloning Man is bad, but what about Man's Best Friend? Most people aren't bothered as much—or at all—by cloning dogs, compared with cloning humans. Reflecting on the reasons for this is very illuminating. The simplest explanation I can come up with is that our concept of people—especially ourselves—is closely linked to the concept of uniqueness, while our concept of a good canine companion does not depend on uniqueness—at least not to the same degree. ("Comments" 21 Feb. 1998)

Hawthorne here responds directly to the reader's observation that, by focusing on "America's favorite pet, the dog," the Missyplicity Project toes the line between offensive (to him, human) cloning and mainstream (livestock) genetic experimentation involving animals. Between the lines, there lurks an acknowledgment of Missy's socially transformative potential in relation to species status: the cloned dog possibly serves a transitional role by publicly mediating popular conceptions of animal abjection and human identity. But Hawthorne's reply hedges this potential as its gist echoes the message of the MissyTale "A Breed Apart," namely that Missy's unique qualities add up to something "good" in terms of companion-animal breeds, but not to a distinct identity comparable with that of the human. Thus canine non-identity, not the supposedly singular identity of the celebrated mongrel, lies at the heart of Missyplicity.

In this respect, the site's narratives replicate on a microcosmic scale the erasure of the mongrel in the history of dog-breeding narratives. Traditionally the sign of canine self-selected breeding, the mongrel dog's body gains new interest in the context of this scientific cloning project as a peculiar genetic text to be decoded and reprinted. But in the process the copied mutt founds a "breed apart" and, instead of threatening the end of the history of dog breeding, she signals the end of mongrel history and, by extension, canine-determined sexuality if not sociality. In other words, the Missyplicity Project's official narratives, far from celebrating the mongrel, work to eradicate the dog-directed conditions of breeding for which it conventionally stands. Consequently, the website, by reinforcing the idea that cloning will serve only human desires in dog breeding, extends the eradication of the mongrel through its adaptation of cloning to the histories of dog breeding.

For hundreds of years human interventions into canine reproductive patterns have involved the human selection of canine sexual partners as well as the continuous documentation—through stud books, paintings, memoirs, and, more recently, photographs, films, and websites—of these couplings and their products. Breeding records, as the necessary evidence of human control required of all registered members of canine breeds, legitimate not only an interspecific (or cross-species) sphere of intimacy but also a hierarchy projected by the human onto canine social spheres. This hierarchy primarily serves to distinguish the breed dog as an emblem of human-canine breeding relationships from the mongrel, whose lack of documented parental matching confirms her inability to meet the otherwise visual "qualifications" of a "blood-dog" (Mann 216). Textualization legitimates canine "marriages" (or sex acts) that are "churched" (or approved by human breeders) only when they conform and contribute to the record of an established breed (Ackerley 60). Within this system, human textual standards determine what counts as canine sexual agency, aesthetics, and domestication, strictly containing cultural forms that otherwise might be destabilized at this cross-species primal scene.

Whereas, from the standpoint of biology, the dog always comes before the human standardization of breed, these terms are reversed from the standpoint of the human dog breeder, for whom the breeding chart always comes before the breed-dog subject. To be well-born, according to this human aesthetic of canines, a dog must bear the visible points or characteristics of the breed in addition to having been born of "well-bred" parents, that is, of a couple who are not only both members of the same established breed but also selected for mating in accordance with human aesthetic standards. Like the valuable livestock being genetically manipulated in the same research compound as Missy's clones, the breed-dog subject thereby serves as a container of peculiarly mixed genetic and aesthetic information: at the end of a line-bred family of dogs, the breed representative inhabits a body structured by the documentation of human-canine interactions as much as by human-regulated intracanine sexual activity.

The creation of canine mongrels remains primarily the provenance of dogs themselves, so that, rather than a potential source of renewal for human dog breeding, the canine mongrel remains an image of "bad" breeding, the limit of which is canine self-selected reproduction. As the MissyTales suggest, the mongrel does not inherently stake out an isolated or separate sphere of canine culture that is opposed to the human-centered culture of breed-dog production. On the contrary, these dog-centered social practices are not distinguished by escape from human involvement so much as they, by definition, confound the anthropocentric terms of human-dog-breeding textualization. Particularly through the euphemism "random bred," human bafflement at mongrel breeding betrays a strategic encroachment on canine sexual agency in breed-dog production; the human dog breeder, positioning canine sexual chaos as the alternative to human choices, rules out the possibility that canine sexual actors can be guided by non-human aesthetics or any other manifestation of non-human animal cultures.

A genre that plays with these cross-species and textual mediations, the dog-breeding narrative historically couples the singular bonds of dog love with "good breeding," a eugenic ideal that, in the context of dog breeding, elides demarcations of human class difference, pet ownership, and the social standardization of visually consistent, highly salable breed dogs. Visual consistency plays a pivotal role in securing the human-dog breeding relationship, textually inscribed in the progression of dog-breeding narratives from the older tradition of canine portraiture. As Harriet Ritvo elaborates, the plasticity of breed as a measure of human control allows owners to assert social status through individual dogs only to underscore the constructedness of breed and, by association, human social class categories (93). Recorded in images and stories, this representational movement indicates both the metaphorical utility of breed-dog images in equating individual breed dogs with their wealthy owners and the profound anxieties about the ways in which metonymical instabilities become visible in cross-species relationships.

Images thus contain the dog breeder's increasingly conflicted sense of identification through the animal, such that the breed animal itself comes to serve human breeders as "an index of their paradoxical willingness aggressively to reconceive and refashion the social order in which they coveted a stable place" (Ritvo 115). Highlighting this contrast, the breed dog's image generally figures the establishment of class through breeding, in contrast to the story of the dog's ownership, which emphasizes the socially revolutionary potential of breeding. The central role of illustrations on the Missyplicity Project's website models the ways in which contradictory desires for social transformation and the fixity of an individual's elite status, troubling the dog-breeding narrative, resolve in the representation of the dog's physical self.

The images thus stabilize the stories' development of this individual into an avatar of breed in a reification process involving the transformation of this invaluable pet into a rarified commodity. Like the most expensive of all breed canines, the show-quality dog, Missy becomes through this use of visual narrative "a special kind of body traced into being by the elaborate procedures" structuring her existence (Watkins 140). The value of such a dog, as Evan Watkins argues, involves not simply her textualization (as, for instance, her place in a breeding record) but, just as importantly, her performance as a spectacular self. For such a dog operates as a "positional good," whose value as a scarce and "elaborately individua[ted]" commodity is determined both by her breed membership and her rarity or, more appropriately, her lack of a "precise exchange equivalent" (145–46). This highly localized valuation of breed may once have been precisely geared to devalue mutts like Missy, who, after her cloning, threatens a radical economic shift through the serialization of the replicated breed dog. In this way, Missy becomes a figure of reification in Fredric Jameson's dual sense, both of the transformation of dog love into clones and, more abstractly, of the effacement of the traces of the cloning process from the cloned products (314–15).

Enabled by illustration with images of Missy, the MissyTales use familiar rhetorical strategies to steep their narrative of dog cloning in the history of human-canine domestication long before Missy's actual clones force the issue, raising questions about how the singular image of Missy comes to stand for the as yet immaterial faces of her clones. From the chaos of mongrel origins, humans select and replicate Missy, the non-sexual, scientific, and dog-love object, who embodies the simultaneous installation of a new breed and a new dog-breeding order through cloning. Positioning Missy as a prototype of a mechanically manipulated dog-breeding process that in turn collapses replicated breed dog with sterilized mongrel dog, the narratives of Missyplicity rely on the replication of images of its central clone donor in order to manifest the rules and language of cultural order in a process akin to what Judith Roof terms "reproductions of reproduction." Whereas Roof's theory refers to an intrahuman pattern of texts that literalize the father in order to assert the "centrality of the paternal" at the expense of the maternal body (14), Missyplicity figures an emergent pattern of what I term "replications of replication" through animal-breeding narratives. This formation is apparent in the rhetoric of the MissyTales, where the human mother strives to literalize a "breed" in Missy's mongrel body, but its resolution requires their linkage with multiple visual images of Missy.

To replicate the concept of replication, the website requires visual confirmation of the human mother's sentiment, so that imaging proves the determining factor, moving Missy from the passive and limited roles of love object and tissue donor to omnipresent spokesmodel for dog cloning. Her visual permeation of the website, in its framing, wallpaper, and foregrounded illustrations, covers over the textual absence of Missy as a social agent. The images of consistently healthy, well-groomed, and smiling Missy have not changed since the site's inception. With her future accounted for by clones and her past by her human mother, these images, as replications of Missy's replication, stand as a bulwark against the historicity of cloning.

Reading this visual dynamic in terms of the replication of replication underscores not only the transformation of Missy's social relationships into her genetic material but also the farther-reaching effacement of the multiple cultural contexts involved in the process of dog cloning. Again, dog love provides a productive confusion of canine agency, as Hawthorne insists that the Project's primary goal of cloning Missy subordinates scientists' interests to her owners' desires, which, as the MissyTales insist, come directly from Missy herself. Clarifying the distinction between what it means "to reproduce" and "to create a genetic duplicate" outlined on the "Welcome" page, Hawthorne elaborates the owners' decision to clone in a way that makes it appear to be instigated by Missy's material qualities:

> [H]er owners . . . believe that Missy's genetic gifts—governing both her appearance and behavior (to some unknown extent)—are significant, and that another dog with the same genetic endowment will probably be a great dog too—though certainly not the same dog. When Missy's owners look at the clones, their appearance will not serve to perpetuate the illusion that Missy lives on, but rather will remind them of the source, the irreplaceable Missy herself. ("Comments" 12 Jan. 1998)

In this version, Missy's clones bear the "genetic gifts" that become her bottom life through her owners' transplantation of them into their duplicate pets. Elsewhere, Hawthorne indicates how the cloning of Missy involves the transformation of pet owning into pets as things by describing her cloning in terms of replacement:

> Other people are already hard at work on countless cloning projects—mostly involving the subjugation of animals. This project projects a unique message: The individual animal called Missy is so dear to her owner that he . . . is driven to replace her by a technological process. ("Comments" 24 Aug. 1997)

While the male owner gains some responsibility for the decision to clone here, the phrasing also suggests that Missy's materiality, her endearing qualities, force him to do it. These passages illustrate how, instead of rising to the rhetorical challenge of developing a language adequate to the proliferation of biologic, genetic, and social contributors to the cloning process, the site's narratives founder on the restrictions and conflicts built into the dog breeders' rhetorics of commodification. In this respect, the passages echo the description in the Missy-Tale "How Missy Found a Home" of Missy's initial procurement—"We decided to try to find another dog—to cheer [the coydog], to make her life more interesting, whether or not she approved"—in that it indicates how dogs' desires have no place in a cloning process designed to serve first and foremost the desires of humans. Rather than imagining a potential conflict of interests across species lines, the narrators of the Missyplicity Project are preoccupied by human opposition to this ultimately human-centered goal of cloning Missy, and they combat what appears to them to be the more formidable threat of public outrage with their stories and images of replicating a "perfect" bitch.

While the images successfully connect Missy to a context of comfort and privilege, they also express anxiety about the loss of human control through the dog cloning process. Among the thirteen images figured in MissyMedia, one indirectly figures Missy as the subject of dog cloning. "Venus de Missy," an animated GIF (sequential still photos layered and syncopated to form an "animated" loop), singles out Missy to provide an image of spontaneous canine generation. Rather than the *Venus de Milo* image cited by the title, this image-sequence activates as it cites Botticelli's *Birth of Venus*. The loop moves Missy from the ocean to land, a movement underscored in the text description: "In this series, Missy surfs a wave, then gives herself a good shake." The absence of humans and other dogs in this image indicates that this "birthing" scene is self-directed and self-contained. It figures Missy as replicating herself in an endless loop, in accordance with art-historical ideals of immortals and scientific fantasies of clones alike.

In contrast, the experimental process of cloning Missy involves several human laboratory workers as well as dogs, in this case a colony of about sixty bitches procured for the Project "from a facility for the breeding of dogs for laboratory research" ("Adoption Center"). These bitches provide two crucial biological components for Missy's clones: eggs out of which their own DNA is evacuated and replaced by Missy's, and wombs into which the cloned embryos of Missy are surgically placed and, ideally, incubated. In return for these services, the Project offers these dogs what no scientific research project to date has offered lab dogs: the chance to be adopted as pets. Toward this end, the predominantly beagle bitches are groomed for another life, offered through the "Adoption Center" section of the Project's website. The ironies of this situation are not lost on one recent visitor, nor, it seems, on his host, Hawthorne:

> Like their yelping brethren in the [experimentation animal] sheds [at TAMU], the Missyplicity dogs were bred for experimentation. But unlike the regular lab dogs, which live out their lives inside cages, the Missyplicity dogs have names and an hour of daily exercise. Here on the gravel, they gallop around their jumpsuited obedience trainer, who housebreaks them for home adoption after their eight-month stay.
>
> "Ours are the lucky ones," Hawthorne says, scratching a beagle's tattooed ear. "Sometimes, walking through here, I feel like Schindler." (Graeber 228)

If the analogy seems confused (is he saying that the Project's scientists and Missy's owners are Nazis?), its context only compounds this confusion. This account carefully notes that, as Hawthorne speaks, "several confused beagles hump his leg," a detail that both highlights the desperate living conditions of this canine harem and echoes one of the more memorable images of Missy herself.

The only MissyMedia image that shows Missy with humans, the color still "Humpty-Doggy" (fig. 10.2) offers a striking image of Missy directing a very canine and possibly sexual gesture toward her human mother's leg. The caption explains, "Although Missy is a spayed female, she sometimes attempts to mount other dogs and the legs of people." The copy goes on speciously to distinguish Missy's similar expressions in intracanine contexts, in which they are "dominance-oriented," from canine-human contexts, in which these gestures express "love and excitement". Missy apparently does not "love" the dogs whom she mounts, only the humans. The anonymous author of this passage surprisingly evokes canine desire in a passage akin to J. R. Ackerley's description of a similar scene as typical of his own bitch in heat. Struggling to record without reducing canine behavior to human terms, Ackerley describes this gesture as always involving a dog feeling "sweet" toward the human, canine, or inanimate object on which it is "bestowed" (63–64). But Ackerley's projection of a common motive to account for the varied objects involved in this typical canine activity stands in stark opposition to the website's account of this parallel image of Missy, which assumes that the bitch's ability to recognize species difference is identical to humans'. With this insistence on the species-specificity of Missy's feelings for (dominated) dogs and (loved) humans perplexedly expressed through the same physical gesture, the Missyplicity Project in this exceptional moment separates Missy's canine and human cultural contexts and quietly reveals how it depends on a conception of human-canine sociality positioned as dominating, if not hostile to, canine social systems.

In this context, the most poignant image perhaps in the MissyMedia archive is the only one that shows her with other dogs. "Friends" (fig. 10.3) shows Missy sniffing two nearly identical Jack Russell terriers in greeting postures. Where all of the other images on the site position her alone or, in the one instance, with her human mother, this single color still envisions her life as a dog among dogs. The image points to a world of intracanine sociality otherwise invisible within Missyplicity and perhaps made obsolete by the Project. The positioning of these breed dogs with Missy aims to prove, as the copy asserts, that "Missy is a social animal," with many human and dog "friends." But it also suggests that, with the advent of dog cloning, these animals can relate to one another only as friends, never lovers, and consequently must look identical to each other as well as to the human breed ideals.

While the technology of cloning promises to perfect the process of replicat-

---

10.2. (*above right*) "Humpty-Doggy."

10.3. (*below right*) "Friends."

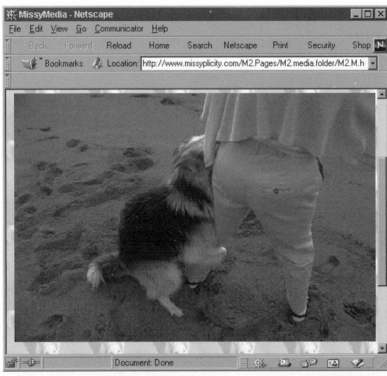

ing dog love as it is filtered through the narrative and visual structures of breed-dog replication, the Missyplicity Project website suggests that it can also be used to challenge the dog-breeding narrative tradition by treating the spayed mongrel bitch's ability to self-reproduce as a problem rather than a solution. Yet its presentation of Missy as a model dog, vessel of the indeterminate yet nonetheless "perfect breed," beloved beauty incapable of reproduction, as well as favored pet of "her humans," exacerbates the unevenness of human-animal relationships underlying the special conditions of this mongrel's replication. The website labors to relieve dog cloning of the stigma of bad breeding but, by positioning replication as the inevitable fruition of dog love, ultimately amplifies rather than resolves the problems of human intervention into canine reproduction.

With these uses of images and narratives, the Missyplicity Project thereby brings together the contested sites of "paternity, maternity, or 'clonerity'" (Elster F50) in a representation of replicative anxiety akin to the "symptomatic representations of reproductive anxiety" that Roof argues mark human-specific texts at the end of the twentieth century (30). Interlacing the rhetorics and images of dog loves and breeds, the site suggests that dog cloning, particularly as it is used to validate human breed aesthetics, imposes limits on the range of acceptable canine behaviors as well as bodily forms. But these boundaries, which may eventually rest on Missy's yet embryonic clones—cocooned within this web of biogenetic narratives, behavioral standards, and the sites of their construction—for the time being end as they begin in images and stories of Missy herself.

Given this import, MissyTales and MissyMedia prove so powerful that they come to supercede the incorporation of feedback in the Project's ongoing self-promotion. The "bi-directional" interactions on the Missyplicity website, spanning the first two years of the cloning project's development, constitute a short narrative interplay between readers and designers of the Project's website that ends abruptly upon the announcement of the incorporation of Genetic Savings & Clone and the establishment of its promotional website, *Savingsandclone.com,* multiply linked through the text of the Missyplicity "Welcome" page and banners on all other pages of the site. The readers' postings to the Missyplicity Project website include several requests to clone specific pets, justifying the development of the subsidiary company, if not the incorporation of these bi-directional narratives into what then becomes a linear story.[3] The creation of the "Forums" section of the GSC site, the first posting to which is dated two years after the last reader's posting to the "Comments" section of the Missyplicity site, develops the bi-directional narrative from the research sector and reestablishes it in the commercial sphere. In addition to Hawthorne's direct responses to readers' postings, what remains prominently consistent between these two websites is the use of images and stories of Missy as structuring devices for the presentation and proliferation of dog cloning. Missy's digital presence (again, a combination of images and stories) welcomes visitors not only to Missyplicity but also, posed alongside Hawthorne, to the Pet Division of GSC (fig. 10.4).[4]

What these proliferating visual mediations of Missy and the marketing of

10.4. Man and Dog. Genetic Savings & Clone.

her cloning suggest are the ways in which, as E. Ann Kaplan and Susan Squier argue, the technologies of "cloning . . . and the proliferation of images of a clone are linked," not only in "the analogy between reproduction at the bio-medical level and that at the photographic or digital level," but also by their parallel contributions to a more "general movement . . . to take biological and intellectual properties out of the public domain and hold them for private industry" (5). In other words, it is possible to purchase the replication, through the body of another, of the individual whose image is replicated, in this case a cherished pet and now "celebrity" dog, but only at the cost of extending the process of replication and accelerating the global human culture of capitalism. What gets lost amid these replicated images of the elaborately individuated replicant is the history of its production through what Mary Midgley calls "mixed communities," in this case, of clone, human, and non-human animal cultures. Missy's visual appeal works in conjunction with her ardent acquaintances' testimonies to sell an idea of the privatization of biological research, namely, cloning on demand, that diminishes the range of cultural diversity even before its products threaten to limit the spectrum of biodiversity.

This dynamic of visual storytelling in the cultural production of clones underscores, however, that the social viability of cloned animals relies on specific uses of visual imaging technologies, so that their future public visibility remains uncertain. Posting on the eve of his transformation into both Missyplicity Project Coordinator and Genetic Savings & Clone CEO, Hawthorne makes no bones about relying on visualization technologies to foster widespread support of cloning, as he asserts in reply to a reader's posting, "It will take the general public about five minutes to accept dog cloning, once the first puppies are born and shown on the evening news" ("Comments" 13 Jan. 1998). Hawthorne assumes that, more than even Missy's images on the site, the televised images of her clones as pups (like those of Dolly as a lamb) will popularize pet cloning. Even before these dog clones have been successfully whelped, however, the Missyplicity Project has begun this visualization process. Consequently, Missy's embryonic clones slouch toward this electronic Bethlehem to be born in the public eye under the shadow of dog love, breed narratives, and the image of "a really great mutt" ("Welcome").

Notes

1. Reporters at the *Dallas Morning News* "examined tax records and California dog tags" to determine that one of Missy's owners is John Sperling, early proponent of the failed Biosphere 2 and founder of the for-profit University of Phoenix. Sperling, however, denies ownership of Missy (Cohen). The Missyplicity Project is contractually obliged to maintain the owners' anonymity.

2. Here I borrow Donna J. Haraway's mutation of Bruno Latour's concept of "technoscience" to designate "a form of life, a practice, a culture, [and] a generative matrix" (Haraway 50). Elsewhere Latour indirectly indicates that such a formation may not be tenable in relation to "Western" science, within which cultural mediation becomes conceivable only on the condition of its denial (89).

3. In the overview page of the "Pet Division" of GSC, Lou Hawthorne boasts, "[A]fter receiving hundreds of requests, we're now making our Missyplicity expertise available to you and your pets."

4. "Clone Missy" and "Develop relatively low-cost commercial dog-cloning services for the general public" are, respectively, the first and last entries in the "Project Goals" page of the Missyplicity Project website. An addendum to the "Pet Division" overview page of the Genetic Savings & Clone site indicates how this offshoot project attempts to simultaneously achieve the latter goal and secure a marketing edge: "The cost of gene banking at GSC is fully deductible from the future cost of cloning at GSC—something other gene banks can't offer!"

Bibliography

Ackerley, J. R. *My Dog Tulip*. 1956. New York: Poseidon, 1987.

"Adoption Center." *The Missyplicity Project*. Accessed 29 Jan. 2002 <http://www.missyplicity.com/M2.Pages/M2.adoption.folder/M2.adoption.html>.

"A Breed Apart." *The Missyplicity Project*. Accessed 29 Jan. 2002 <http://www.missyplicity.com/M2.Pages/M2.tales.folder/M2.taleBreed.html>.

Cohen, Hal. "Field Notes: Bubble Trouble." *Lingua Franca* 8.8 (1998): 14.

"Comments." *The Missyplicity Project*. Accessed 29 Jan. 2002 <http://www.missyplicity.com/M2.Pages/M2.comments.html>.

Doane, Mary Ann. *The Desire to Desire: The Woman's Film of the 1940s*. Bloomington: Indiana University Press, 1987.

Elster, Nanette R. "Who Is the Parent in Cloning?" *Cloning Human Beings: Commissioned Papers*. National Bioethics Advisory Commission. Rockville, Md.: GPO, 1997. F41–50.

"Frequently Asked Questions." *The Missyplicity Project*. Accessed 29 Jan. 2002 <http://www.missyplicity.com/M2.Pages/M2.faqs.html>.

Garber, Marjorie. *Dog Love*. New York: Simon and Schuster, 1996.

Genetic Savings & Clone. Accessed 29 Jan. 2002 <http://www.savingsandclone.com/>.

Graeber, Charles. "How Much Is That Doggy in the Vitro?" *Wired* 8.3 (2000): 220–29. Also on the Web at <http://www.wired.com/wired/archive/8.03/dog.html>, accessed 29 Jan. 2002.

Haraway, Donna J. *Modest_Witness@Second_Millennium.FemaleMan©_Meets_Onco-Mouse™: Feminism and Technoscience*. New York: Routledge, 1997.

Hearne, Vicki. *Adam's Task: Calling Animals by Name*. New York: Knopf, 1986.

"How Missy Found a Home." *The Missyplicity Project*. Accessed 29 Jan. 2002 <http://www.missyplicity.com/M2.Pages/M2.tales.folder/M2.taleHome.html>.

Jameson, Fredric. *Postmodernism: or, The Cultural Logic of Late Capitalism*. Durham: Duke University Press, 1991.

Kaplan, E. Ann, and Susan Squier. Introduction to *Playing Dolly: Technocultural Formations, Fantasies, and Fictions of Assisted Reproduction*. Ed. E. Ann Kaplan and Susan Squier. New Brunswick: Rutgers University Press, 1999. 1–13.

Latour, Bruno. *We Have Never Been Modern*. 1991. Trans. Catherine Porter. Cambridge, Mass.: Harvard University Press, 1993.

Levin, Ira. *The Boys from Brazil*. New York: Random, 1976.

Mann, Thomas. "A Man and His Dog." 1930. *Death in Venice and Seven Other Stories*. Trans. H. T. Lowe-Porter. New York: BOMC, 1993. 215–88.

Midgley, Mary. *Animals and Why They Matter*. Athens: University of Georgia Press, 1983.

"MissyMedia." *The Missyplicity Project*. Accessed 29 Jan. 2002 <http://www.missyplicity.com/M2.Pages/M2.media.html>.

The Missyplicity Project. 15 Aug. 1997. Texas Agricultural Experiment Station, Texas A&M University. Accessed 28 Jan. 2002 <http://www.missyplicity.com/>.

"Pet Division." *Genetic Savings & Clone*. Accessed 29 Jan. 2002 <http://www.savingsandclone.com/overview.cfm?div=Pets>.

"Press Releases." *The Missyplicity Project*. 24 Aug. 1998. Accessed 29 Jan. 2002 <http://www.missyplicity.com/M2.Pages/M2.press.releases.html>.

"Project Goals." *The Missyplicity Project.* Accessed 29 Jan. 2002
    <http://www.missyplicity.com/M2.Pages/M2.goals.html>.
Ritvo, Harriet. *The Animal Estate: The English and Other Creatures in the Victorian
    Age.* Cambridge, Mass.: Harvard University Press, 1987.
Roof, Judith. *Reproductions of Reproduction: Imaging Symbolic Change.* New York:
    Routledge, 1996.
Sirius, R. U. [pseud.]. "Cloning the Pooch." *Salon* 29 Mar. 1999. Accessed 29 Jan. 2002
    <http://www.salon.com/21st/feature/1999/03/29feature.html>.
Turner, Stephanie. "More than Genes." Online posting. 28 Mar. 2000. Genetic Savings
    & Clone Forums. *Genetic Savings & Clone.* Accessed 29 Jan. 2002
    <http://www.savingsandclone.com/forums/
    Thread.cfm?CFApp=2&Thread_ID=16&mc=4>.
Watkins, Evan. "Your Dog's Just a Dog: Literary Scholarship and Market Politics."
    Everyday Exchanges: Marketwork and Capitalist Common Sense. Stanford:
    Stanford University Press, 1998. 128–59.
"Welcome." *The Missyplicity Project.* 8 Feb. 2000. Accessed 29 Jan. 2002
    <http://www.missyplicity.com/>.
Wright, Susan. "Recombinant DNA Technology and Its Social Transformation." *Osiris*
    2nd ser. 2 (1986): 303–60.

11 Immersed with Animals

Nigel Rothfels

The Bronx Zoo's virtual tour of the Congo Gorilla Forest exhibit begins with a closeup shot of a young gorilla's face (fig. 11.1) with the legend "Imagine being this close!" ("Welcome"). This is a compelling thought, but the promise on the screen is misleading, for, although the exhibit appears to deliver a closeness to "nature," it is clearly not designed to get the visitor "this close" to a gorilla—except perhaps in the virtual space of the Wildlife Conservation Society's website, *wcs.org*. Indeed, the goal of the exhibit seems at least partially to be to hide the animals, and this paradox of an exhibit designed not to exhibit is at the heart of most of our "better" contemporary zoos.[1]

The virtual tour of the Congo Gorilla Forest guides one along the trails of the new exhibit, and around every turn—or after every click of the mouse—a new and exciting vista opens where one should expect to see the unexpected. Approaching the Okapi forest through the heart of a hollowed-out "tree" (fig. 11.2), for example, our virtual tour-guide explains,

> You emerge from the fallen tree and see a waterfall. A Goliath frog rests in the spray waiting for his unsuspecting dinner to fly by.
> Across the path a stream flows over the remains of a half eaten fish. The plants on the other side of the stream move, but you still can't see. . . . Wait! There! Slowly, phantom-like, it becomes visible. An okapi! As you search the forest for signs of others, a Congo peacock meanders by.
> You want to stay and watch the okapi but something drives you forward . . . maybe there are others just around the next turn. ("Okapi")

The actual Congo Gorilla Forest exhibit (as opposed to its virtual representation) is an example of what zoo designers have called an "immersion exhibit," a place where both the animal and, increasingly, its human observer appear to be "immersed" in a natural environment. In some cases these exhibits are quite large. According to the Wildlife Conservation Society, for example, the "Congo" comprises 6.5 acres and contains representatives of 75 animal species (including 22 of the namesake gorillas), "15,000 living plants of more than 400 species," and "ten miles of fabricated vines, great fabricated trees (epoxy, steel and urethane), replicas of giant *Ceiba* trees, stilt rooted *Uapacas*, [and] trees damaged by elephants." "The Congo," the website declares with little sense of irony, "holds the distinction of being the largest African rain forest ever built" ("Fast").

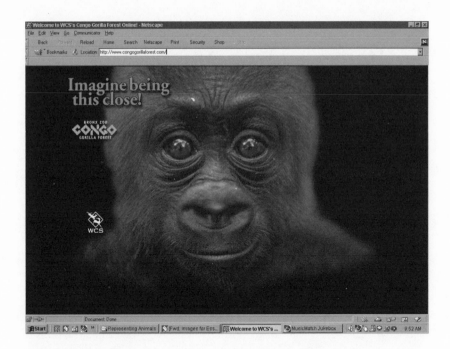

11.1. Welcoming screen of the virtual tour of the Congo Gorilla Forest.
Screen capture.
Photo: D. DeMello. © Wildlife Conservation Society.

11.2. Exiting the "Walk-Thru Tree," Congo Gorilla Forest.
Photo: D. DeMello. © Wildlife Conservation Society.

11.3. Butterfly Garden.
Courtesy of the Milwaukee Public Museum.

In other cases, however, the exhibits occupy smaller and more intimate spaces, such as the currently highly popular butterfly exhibits where visitors walk gently through Edenic gardens while hundreds of butterflies land softly on flowers, feeders, and outstretched hands (fig. 11.3). Whether large or small, though, the new immersion exhibits have become the ambition of practically all contemporary zoological gardens. Annually, in fact, American zoos are spending millions of dollars to construct these new environments designed to transport human visitors to faraway, mysterious, and "wild" places. Serious and ambitious zoos of today, we are told, need new and exciting kinds of exhibits, and the new immersion exhibits fit the bill.

To be sure, these exhibits did not spring up out of nowhere. In the 1950s, '60s, and '70s, most zoological gardens in Europe and North America expanded on experiments begun earlier in the century to convince zoo audiences that animals might profitably be shown in something resembling "natural" environments. In some cases "nature" was to be understood in only the most abstract sense. Projects like the London Zoo's Elephant and Rhino Pavilion of 1965, for example, seem to be meant to appear "natural" without resembling anything that could be seen in nature. From the outside, people stood before a low wall with a chest-high bar and looked across a small moat at the conglomerate form of the central pavilion—an imposing, light-earth-toned, rough concrete building which has been described as "zoomorphic New Brutalism, marvellously expressive of its inhabitants" (Guillery 43). The building is an abstracted "nature," and even though by the 1990s the barren spaces inside the building had been softened and jungle-ified through the addition of supplementary lighting and plantwork, the building still carries a highly intellectualized ambience. Abstraction, though, was really only one approach to "creating nature" in zoos. Not surprisingly, many zoos opted for the more economical option of removing bars, adding glass, and hiring artists to create sometimes astonishing, if also occasionally bizarre, backdrops designed to meet the expectations of the human audience (figs. 11.4 and 11.5).

Today, however, the sheer scale and sophistication of the immersion exhibits are making these older displays seem as antiquated to us today as late-nineteenth-century exhibits must have seemed in the 1950s. Most impressively, it all seems so self-evidently "correct." Isn't it obvious that we should use our immense technological means to create these stunning environments for the benefit of both the animals and their audiences? After all, the concept is so sensibly simple. As one author notes enthusiastically, "Add real plants, real soil, water, sound effects and a touch of imagination and you have an immersion exhibit" (Koebner 84). Even if it is unclear whether a tree made out of concrete and epoxy really is a tree as far as an opossum or a bonobo is concerned, wouldn't such a tree be necessarily more interesting to any animal than steel bars and an unnatural-looking plastic ball? Most importantly, wouldn't such an exhibit meaningfully enhance a zoo-goer's experience and make him or her more sensitive to the environmental issues confronting animals in the wild? At "the heart of the forest" for school groups visiting the "Congo" in the Bronx, for example, are both the Charles Hayden Foundation Treetop Lab (with special views into an aviary, a guenon monkey exhibit, and the "Judy and Michael Steinhardt Mandrill Forest" exhibit) and the Bodman Foundation Congo Lab, which "overlooks gorilla habitats with very special encounters" ("Fast"). In this setting, it is argued, children and adults can truly learn about the environments in which animals live and, in the case of the "Congo," actually become involved in conservation efforts as they designate to which *in situ* research project they would like their entry fee devoted. In short, aren't the rationales for such exhibits as thoroughly convincing as the exhibits themselves (fig. 11.6)?

In the new, more perfect world of the immersion exhibit, a better "nature" is created for animals: food is plentiful and more and more interesting; parasites are carefully managed; sicknesses are combated with the full range of modern medical technologies; climate is thoroughly regulated by advanced computer systems; human visitors are obscured behind naturalistic banks of vegetation; sounds of better-than-real forests, seashores, and mountain escarpments are piped in through camouflaged speakers; and successful propagation is the clear measure of happiness and health. Indeed, despite the fact that the claims of the new exhibits are sometimes overinflated, the general consensus seems to be that our most advanced zoo immersion exhibits are significantly different from their

11.4. Jacqueline Hayden, *The Lion in Winter.*
Courtesy of the artist.

11.5. Frank Noelker, "Untitled," from Zoo Pictures. 1997. Iris print.
Courtesy of the artist.

11.6. The Congo in the Bronx.
Photo: D. Shapiro. © Wildlife Conservation Society.

nineteenth- and early-twentieth-century precursors. But what precisely is the nature of that difference? To answer that question, it is helpful to look briefly at the differences between the London Zoo in the mid nineteenth century and the early-twentieth-century Hagenbeck's Animal Park (*Tierpark*) in Stellingen, a suburb of Hamburg in Germany. For it is in the contrast between these two remarkable places—the London Zoo being perhaps the most admired in the nineteenth century and the Hagenbeck Park being widely regarded as the birthplace of the twentieth-century zoo—that one can see the origins of our contemporary immersion exhibits.

In her "Memoir of Sir Thomas Stamford Raffles, F.R.S.," Raffles's widow recalled that around 1817 "he meditated the establishment of a society on the principle of the *Jardin des Plantes*, which finally, on his last return from the East, he succeeded in forming, in 1826, under the title of the Zoological Society of London" (qtd. in Scherren 7). Raffles's primary aim was to create a forum in which those interested in specifically zoological topics could study and present scientific papers. Raffles and others insisted that the new Society was needed because the Linnaean Society, established in 1777 to cultivate the general study of natural history, had focused too narrowly on botanical studies. In response, the broad objectives of Raffles's organization were to advance zoological science in its aspects of classification and description, and to domesticate new animals to the uses of man. The Society's initial goals, therefore, were the "formation of a collection of living animals; a museum of preserved animals, with a collection of comparative anatomy; and a library connected with the subject" (Scherren 20).

With the founding of the Zoological Gardens in Regent's Park in 1828, however, the collection of animals swiftly assumed an added character that was anything but scientific. Raffles's early proposals included the idea that the zoological collection should also both "interest and amuse the public" (Scherren 7). In the end, this latter quality would become perhaps the most important mandate behind the development of the Gardens throughout the century. Indeed, by the middle of the nineteenth century, the Gardens resembled a public place of entertainment much more than a scientific station. While originally only Fellows of the Society and their guests were admitted to the Gardens, within a dozen years the general public was admitted on Mondays and Tuesdays for the price of a shilling each, and on other days with payment and a written voucher from a Fellow. By the end of the 1840s, the public was admitted Monday through Saturday, paying sixpence on Monday and a shilling the rest of the week; children paid sixpence all week. By 1850, however, the last social barriers began to crumble and even the Promenade Days, which had been set aside especially for the Fellows, had also been taken over by the general public and had become part of the regular Saturday fanfare at the zoo. They had, indeed, become part of "general admission." Thus, by the second half of the century, the Gardens had become a well-established and highly acceptable venue of outdoor public entertainment—something, it seems, between an urban nature park and an

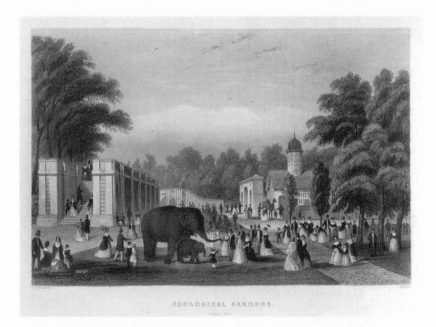

ZOOLOGICAL GARDENS.

11.7. Le Pettit (after Thomas Hosmer Shepherd), "Zoological Gardens, Regent's Park."
From *Views of Mighty London: Its Environs and Royal Palaces* (London, 1854).
Yale Center for British Art, Paul Mellon Collection.

amusement park—and this development seems to have been typical of the other major zoological gardens of Europe and the United States.

The fundamental qualities of the London Zoo in the mid nineteenth century are clear in the series of "views" of the Gardens executed by Thomas Hosmer Shepherd. His "View of the Music Lawn" captures the general feeling in these works.[2] At the right rear of the scene, we see the Camel House Clock Tower, designed by the zoo's first architect, Decimus Burton, and built in 1828. Down the left runs the Carnivora Terrace of 1843, with its ionic pilasters and flat surfaces surrounding picture-frame cages housing lions and tigers and bears. At the rear of the scene stands the oblong Polar Bear Cage, constructed around 1832, with its high inward-arching steel bars. But more than the animals at this zoological garden, indeed more than the remarkable buildings, the central focus of the print remains the lawn and its human inhabitants. Indeed, the presence of people seems absolutely essential to this work. Friends, couples, and families with well-behaved children walk and talk, stand before the cages studying animals, gather to socialize, and present edibles to an apparently free-roaming elephant.[3] (fig. 11.7)

This was a place designed by the bourgeoisie for its own education and amusement. The atmosphere of this remarkable public institution thus encouraged social events such as band concerts and promenades, and the presentation of animals in contexts saturated with human references. In a passage which seems in many ways characteristic of the period, for example, a guidebook to the Gardens from the early 1860s describes the path from the main entrance to the Carnivora Terrace:

> From the rustic lodges at the north, or main entrance, runs a broad terrace walk, in a straight line onwards, bordered by flowers, shrubs, and trees on each side, and continued at the same level for some distance, over the lower ground, by a handsome viaduct [the Carnivora Terrace], which covers a long range of roomy cages beneath, and in itself forms one of the most striking objects in the Gardens. On this platform, which is balustraded at the sides, the visitor may pause for a moment, to contemplate the extensive view presented of Regent's Park, and the mighty Metropolis beyond. Save its smoke, however, and the mist, or dense air, perpetually hanging over it but little of the latter is visible. Still it is not less present to the imagination's eye, and the contrast is the stronger when compared with the tranquil scene around.[4] (*Zoological* 5–6)

The passage emphasizes the Gardens as a place of quiet repose in the heart of an industrial city. Underscoring the smoke and thick air seemingly "hanging" over the city, the passage suggests that this air is somehow magically lifted at the Gardens, a place where thoughtful people could find an opportunity to contemplate the striking contrasts between the densely populated city and hushed nature. Of course, the animals could not always be relied upon to cooperate with this idyll. The raucous sounds of the bird houses—made all the worse through their frequent construction as glass conservatories—the smells of the great cats, the inopportune matings, and the sometimes pathetic conditions of the captive animals all drew regular criticism. In a typical letter to a director of one of the bourgeois zoos (in this case William Hornaday of the Bronx Zoo), for example, John P. Haines, president of the New York chapter of the American Society for the Prevention of Cruelty to Animals, wrote with a common lack of irony that the collections should not be in any way depressing, or perhaps unedifying, for the human visitor.

> Dear Sir,
>
> A friend of animals called at this office yesterday, and after highly complimenting the manner in which the animals are housed at the NYZS, said that she had been distressed by one thing she saw, and to which she asked us to call your attention. She said that she noticed in the bird house, in the cage devoted to birds indigenous to this section, a poor robin which was apparently in a droopy and sickly condition. It seemed to our complainant that in view of the prevalence of the common robin, and the ease with which a specimen can be obtained, the society might

at least confine a healthy bird, if it is considered necessary to keep a robin in confinement.

We refer this complaint to you, knowing that you will do what is proper in the premises.

Yours very truly,
JPH
President[5]

Characteristically, the focus of this complaint is not that it might be a good idea to see if someone could do something for the sick robin, but that it is somehow wrong to exhibit a sick robin in a cage. This reasoning stems, of course, from zoos being imagined as places of amusement; there is, in the end, little fun to be had in "drooping" animals. In providing amusement, nothing beat the act of offering food to exotic animals. Indeed, feeding the animals, either by oneself or through a zookeeper, was a central part of visits to the zoo, and the extended arm holding out food to the animals is perhaps the quintessential gesture of these places. This is as clear in Shepherd's "View of the Music Lawn" as it is in innumerable other illustrations depicting zoological gardens of the period.

Throughout the bourgeois zoo, then, we see the overwhelming presence of "the public." Pervading the magnificent buildings (such as Berlin's huge elephant house, built in imitation of a Hindu temple, or London's neo-classical Carnivora Terrace) was a way of envisioning animals in human contexts. The popular and ornate conservatories, with their apparently delicately wrought cages standing in hot humid light and variously housing birds, monkeys, or smaller mammals, thus owed their design almost entirely to cultural expectations which exalted the presence of civilized man in a world of beasts.

If, in the nineteenth century, European and American zoological gardens, with their garden teas and concerts alongside science and education, were clearly designed to consecrate the tasks of enlightened and bourgeois progress in the world, the beginning of the twentieth century saw the narrative structure undergirding zoological gardens undergo a fundamental shift. The key element to that shift appears to be the opening of Carl Hagenbeck's Animal Park, a place which both in its day and ever since has seemed remarkable to almost every historian of zoos.[6] After over half a century of working with exotic animals, Hagenbeck had imagined a new kind of zoo, and at his Animal Park the "immersion exhibit" was first deployed on a large scale. In order to understand just how this form of exhibit represented a new—and somehow postindustrial or postbourgeois—way of viewing the exotic world of animals, a way that today is deeply connected to the virtual worlds made possible by cable and computer, the nature of what is commonly called the "Hagenbeck revolution" must first be clarified.[7]

Carl Hagenbeck's zoo opened in 1907. The foundation of the company, however, dates back to the middle of the nineteenth century, when Hagenbeck's father, a fishmonger in Hamburg, began buying and selling exotic animals arriving at the port city. By the 1870s—and here the fortunes of the company paralleled those of similar firms in Hamburg that concentrated on such other natural resources as guano, sugar, coffee, palm and whale oil, and rice—the business had evolved from a sideline interest of a small fish shop to the world leader in the international trade in exotic animals, a position that remained unchallenged until the beginning of World War I. Zoological gardens, circuses, and private collectors around the world bought their animals from Hagenbeck.

Carl Hagenbeck did not restrict his business simply to trading exotic animals, however. Perhaps most striking, as we look back, was the company's decision in 1874 to begin procuring indigenous people from all over the world for presentation in highly profitable spectacles to European scientific societies and the general public. Then, while continuing his profitable trade in animals and people, Hagenbeck began in the late 1880s to exhibit a series of unique animal acts, the animals for which, he claimed, had been trained in altogether new and humane ways. Finally in 1907 the firm's animal business, exhibitions of people, and performing-animal acts found a permanent home in the new Animal Park, a zoo without the iron bars that had become the most discomforting object among visitors to the older zoological gardens. Based on experiments begun over a decade before, Hagenbeck's Park, with its panoramas in which the animals were separated from each other and the public by carefully hidden moats, became the model zoo for the remainder of the century (fig. 11.8). Here, animals appeared to be living in the wilds of Africa or India even though they actually lived in a zoo in northern Germany. At Hagenbeck's Park visitors could observe "exotic" animals and even peoples in their "native habitats"—the African jungles, Russian steppes, American plains, and Arctic ice—without ever encountering a bar or visible barrier, and without ever leaving the comfort of their own "civilization."

Visiting Hagenbeck's Animal Park in the Hamburg suburb of Stellingen in its first days, Friedrich Katt, a correspondent reporting to the journal of the association of German zoological gardens, noted that Hagenbeck had always "occupied himself with completely different issues from those of the scientifically oriented zoologist who stands at the head of the older zoological gardens." Hagenbeck's past as an animal dealer and trainer, Katt argued, had led to his creating "something at essence popular, an animal show for the visiting public and the animal buyer, something, therefore, totally different from a zoological garden as that concept is generally understood" (371). While admitting that the blatant "theatricality" of the project seemed to "deviate" from the traditional gardens, Katt conceded that "one has nevertheless seen something unusual, something gigantic, when one leaves" and "Hagenbeck's enterprise has assured itself a place in the history of keeping animals as an entirely new kind of zoological institute" (372).[8] Hagenbeck's Park was something different, something more exaggerated and more exciting than any "normal" zoological garden. Recognizing the Park's two distinctive elements of commerce and theater, Katt was overcome with the impressive display and somewhat baffled by it at the same time.

11.8. The Main Panorama at Hagenbeck's *Tierpark* (mid 1920s).
Courtesy *Hagenbecks Tierpark.*

Katt's sense of bewilderment was generally shared by his professional readers, who tended to conclude that Hagenbeck and his new zoo appealed only to the basest interests of the public. Responding to the acclaim that the Park had received in the papers, for example, Kurt Priemel, the director of the zoological gardens in Frankfurt, responded sarcastically,

> Hagenbeck's gardens are described as the "Seventh Wonder of the World," as "The Zoological Garden of the Future"; everything that one sees in Stellingen is supposed to be completely "new and unique," the methods used there for acclimatization and care of exotic animals are supposed to touch on "totally new principles," to have been called into existence entirely to revolutionize zoo keeping, and "unsuspected perspectives" are everywhere supposed to present themselves. (N.p.)

Priemel concluded that newspaper editors "see with the eyes of the great masses, exactly for whose visual desires the Stellingen installations were designed. Of all the beautiful and remarkable things that Stellingen truly offers, the 'great public' sees only the obvious; they . . . stand enraptured before the so-called 'Grazing Animal Enclosure' and are delighted by the 'Lion Grotto' in the background." Despite the exasperation of this representative of the older zoological gardens, however, and despite a preemptive and devastating unofficial boycott of Hagen-beck's animal dealership by the directors of the major German gardens, who were set on stopping the spread of "Hagenbeckism," Hagenbeck's company survived. Moreover, his utopian illusions of freedom for the animals have been emulated by zoological gardens all over the world ever since.[9]

As well as being a technical accomplishment, Hagenbeck's panoramas began to change the way people thought about animal captivity. Hagenbeck's Animal Park presented an innocent and benevolent view of the world, a sort of idealized existence in which the structure of the zoo itself disappeared and the animals lived beside one another in peace. As one of Hagenbeck's assistants put it, Hagenbeck had wanted

> to create an animal paradise which would show animals from all lands and clima-tological zones in a manner suitable to their life conditions, not from behind bars and fences, but in apparent total freedom. This paradise would also exhibit people of all colors. It would be a nature sanctuary in the most truthful sense, a world in miniature; and thousands of visitors would be able to make a danger-free trip around the world and stroll peacefully under palms. (Zukowsky 9)

There was, therefore, more to this illusion than a contrivance which almost invisibly separated one kind of animal from another. To be sure, at the heart of Hagenbeck's illusions was the desire to mask the obvious fact of the animals' captivity. More than anything else, it was the iron bars—marking so clearly the captivity of the animals—which repeatedly caught the attention of visitors to the older zoos. Rainer Maria Rilke, for example, who had visited the *Jardin des Plantes* in 1907, wrote of the panther, "The bars which pass and strike across his gaze / have stunned his sight: the eyes have lost their hold. / To him it seems there are a thousand bars, / a thousand bars and nothing else. No world." For many—perhaps even most—observers at the end of the nineteenth century, the zoo was clearly understood as a place of captivity, a place where animals were locked up. In response to the growing public discomfort with bars on cages, Hagenbeck eliminated the bars. But in so doing, he did more than simply that; indeed, Hagenbeck replaced the bars with narratives of "freedom" and "peace among the animals." Hagenbeck's exhibits, with their "contented people" and "free animals," answered the public's concerns about captivity with a gentle smile. At Hagenbeck's—and now at most modern zoos—the animals were not only not behind bars, they were safe and happy and long-lived.

Already in the very first years of Hagenbeck's Park, the company had begun to modify the original promotions of the Park as a re-creation of Eden or perhaps an inkling of the Kingdom of God, and adopted instead what would become the dominant metaphor for zoos in the twentieth century: the Ark. From a paradise where predator and prey lived side by side in peace, Hagenbeck's Park became a sanctuary from a violent world and even a sanctuary from the brutal realities of the evolutionary "fight for survival" (Zukowsky 58). Surprisingly quickly, it seems, Hagenbeck's Park was transformed into a place where animals, besieged on all sides in the wild, could find refuge in the hands of a congenial old man who became the best friend and perhaps last hope of the animals of the world.[10] As Ludwig Zukowsky, a scientific assistant at the Park, put it in 1929, "in the act of giving his animals, the creatures he loved, a home free of need and misery, Hagenbeck preached that all creatures of the wide, beautiful, roomy Earth had a safe place where they would be secure from the murder and greed of unreasonable and callous people" (61). Pointing to the issue of captivity, another assistant tried to explain the motivations for Hagenbeck's panoramas similarly: "Also in Carl Hagenbeck the wish grew on the basis of his many experiences in caring for and keeping animals—and not least, on the basis of his character as a lover of animals—to offer his animals accommodation as appropriate as possible to their nature, where they could romp to their hearts' content, and thereby overcome to a certain degree the misery of captivity" (Sokolowsky, *Carl Hagenbeck* 43). As Hagenbeck himself wrote, in his park "ibexes, chamois, and antelopes need not trust their lives in captivity to low cages, but rather could strive for the heights on a cliff-like ridge . . . [and the] king of the animals moved about in freedom, in proud majesty in his wide grotto" (*Von Tieren* 176).

The metaphor of the Ark earned the Park, together with almost all zoos in the twentieth century which adopted the idea, a profoundly resonant justification for their continued existence in the face of their critics. Hagenbeck's associates, both during his life and after his death in 1913, in fact, have all but suggested that the very future of life on earth rested on the earnest striving of the animal lover Carl Hagenbeck. While noting the laudable efforts of various conservation societies seeking to protect wildlife in the 1920s, Ludwig Zukowsky, for example, insisted that the only way to prevent the extermination of animals was to teach the masses to love them—and this instruction was provided by both the life of Hagenbeck himself and by the Ark he established to protect them. Zukowsky writes,

> Then comes the great friend of animals, Hagenbeck, and he calls to everyone: come into my beautiful animal park, into my magnificent animal paradise, look at all the diverse creations of God, learn to understand and love them, enjoy them and then go out and protect them across the globe from pursuit and extermination! And the people come in droves, not simply out of curiosity or the desire to see, but also driven by a longing for nature; they feel that they have lost their connection to Nature. When animals can outdo us in the virtues of courage, faith, and patience, when they can be models for us in their love of their offspring, when they return good deeds with thankfulness and trust, they should not be our enemies, but rather must be our friends! (61–62)

Indeed, a visit to Stellingen, we are told, not only promoted the protection of the animal kingdom but also restored the essential humanity of men and women in a rapidly changing and "dehumanizing" modern world.

Nevertheless, however much Hagenbeck and his followers wanted to put a positive spin on the company, it remained difficult to represent an enterprise that thrived on the capture, trade, and exhibition of animals and people as some

11.9. Two youths and young gorilla from Cameroon at Hagenbeck's Park
(1908).
Courtesy *Hagenbecks Tierpark.*

kind of conservation organization and perhaps the last best hope of animals in the world. The difficulty of that challenge, however, seems not to have deterred anyone. Indeed, the company's repeated efforts over the course of the last hundred years to paint the congenial old Hagenbeck as a modern Noah and his Animal Park as, alternately, Eden, Paradise, or the Ark speaks clearly to the ironies inherent in Hagenbeck's diverse and remarkable enterprises.

These ironies are thoroughly embedded in the company's history. Consider, for example, the photograph of a young gorilla and two youths from Cameroon which appeared at the end of Carl Hagenbeck's memoir, *Beasts and Men* (fig. 11.9). According to Hagenbeck, a lieutenant in the German colonial army in Cameroon brought the gorilla to Germany in June 1908 in the company of the two young boys. Hagenbeck writes that the officer had

> hoped to be able to keep this rare animal alive for a long time. Over in Kamerun he had kept it for more than a year, during which time it had enjoyed unbroken health and become a general pet of the station. He hoped to be able to overcome the difficulty of lack of society by providing the two negroes as constant associates for the animal. When the ape first arrived at my animal park he was much weakened with his long sea voyage and took little interest in anything that was going on round about, but he soon picked up, and after a time would sit and walk about on the lawn in company with his two play-fellows, apparently in the best of health and spirits. He had a strong predilection for the petals of roses, and would con-

Immersed with Animals 213

sume large quantities of them. When he had to be taken from one place to another one of the negroes used to carry him on his back, presenting a very droll appearance.[11] (291–92)

It was popularly believed at the time (and is still) that gorillas in captivity died from depression and loneliness more than any other affliction, and so the officer secured two young boys to accompany the gorilla to Europe and to live with it until, presumably, it either died or was sold.[12] The arrangement appears to have seemed perfectly sensible to the officer, perfectly sensible to Hagenbeck, and perhaps even perfectly sensible to the boys themselves. All this sensibility aside, however, the photograph retains a deeply unsettling quality that is only amplified by its caption in the German edition: "*Prophete rechts, Prophete links, das Weltkind in der Mitten*" ("prophets to the right, prophets to the left, the worldling in the middle").[13] Taken from a humorous poem by Goethe commemorating a dinner in 1774 in which he sat between the physiognomist Johann Lavater and the educational reformer Johann Bernhard Basedow, the caption seems meant to add a certain levity to a picture with little obvious humor. According to accounts of the dinner, while Lavater and Basedow carried on at length about their various remarkable ideas, the young Goethe sat quietly and devoted himself to the food—while two thinkers concerned themselves with matters of the mind, a sensualist attended to more immediate concerns.[14]

The awkward relationship between the picture and its caption hints at the more general explanatory dilemmas this picture poses. Indeed, as clear as it seems to many of us that the picture is also about dilemmas of race and empire, power and exploitation, the caption suggests that the photo posed similar problems when it was first taken and published. In the end, the photograph is disturbing, I believe, because in its wearied quiet it is thoroughly receptive to narrative. A photograph like this impels us to imagine stories to explain it, and those stories are the very thing that Hagenbeck's exhibits sought to control. Without the caption, we are let loose to interpret the photograph according to our own sensitivities and sensibilities. With the caption—and its suggestion of a parallel between a young and worldly Goethe and a young and worldly gorilla— the viewer is asked to understand the photograph as somehow amusing. The inherent eloquence in the expressions of the two young boys and the young gorilla are managed and framed by an amusing caption and Hagenbeck's story of their "droll" visit to the Animal Park.

Indeed, "managing eloquence"—attempting to redirect the audience from seeing and imagining an animal's life in captivity—is perhaps the fundamental feature of Hagenbeck's Park. In a photograph of a young elephant before its dead mother we can easily see an often hidden aspect of the animal trade in the late nineteenth century (fig. 11.10); in a photograph of an orangutan with its back to the bars of a cage and its hand clasping a bottle, we can immediately see captivity (fig. 11.11). In our new zoos, on the other hand, with their carefully deployed plants and illusions of freedom which trace back to Hagenbeck's Park, the person poking a rhino with a stick to get it to move is shunned. Now we see

11.10. "Jumbo beside his dead mother" (1908). © Hans Schomburgk— Archive Jutta Niemann.

11.11. "Diogenes" (1908). Courtesy *Hagenbecks Tierpark.*

animals moving quietly in the woods, gathering at a water hole, cresting a verdant hill, and lounging in satisfaction in the afternoon sun on a kopje.

So what precisely was the Hagenbeck revolution—the revolution to which our contemporary zoos consistently trace their origin? The answer that one generally hears is that Carl Hagenbeck invented a way of exhibiting animals by exploiting moats and other techniques that did away with both elaborate buildings and barred cages. This, to my mind, is a small point. Probably every major zoo director at the end of the nineteenth century was aware that bars on cages represented a problem for visitors, and a good many zoos had been experimenting with different kinds of exhibits as a result. Indeed, in the final analysis, and despite the general consensus, Hagenbeck's revolution was not really the moated structures he created. Hagenbeck's revolution was precisely the narratives of freedom and happiness that he developed at his zoo to go along with the newer exhibits. Before Hagenbeck, zoological gardens often struggled to convince the public that it wasn't so bad to be an animal at the zoo; beginning with Hagenbeck, the gardens began finally, and more or less successfully, to renarrate the captive lives of animals. After Hagenbeck, animals were not collected merely for reasons of science or education, or even really for recreation—animals were put in zoos primarily because they were nice, healthy, safe places to be and because the animals were frankly better off there than in the real "wild."[15]

This is the revolution of Hagenbeck, and his legacy is deeply active in the narrative strategies of today's zoos and crops up repeatedly in contemporary descriptions of the purposes of zoos. Rehearsing a largely counterintuitive position, for example, a recent book about zoos claims, "Once, zoos were only for the powerful and the rich, for important guests to visit. Today, the animals in zoos are our important guests" (Koebner 19). No longer the freak pets of a decadent nobility, no longer the victims of imperial contests, animals in at least the major zoos are now, we are to understand, the treasured lucky few. As another proponent puts it,

> A hundred years ago—or even a decade ago in many cases—the life of animals in
> zoos could best be described in the words of Thomas Hobbes: "solitary, poor,
> nasty, brutish, and short." Now, curators of good zoos can effectively guarantee—
> barring hurricanes and other Acts of God—to keep animals alive in captivity in
> most cases for far longer (perhaps several times as long) as they could reasonably
> expect to live in the wild. (Tudge 55)

Freed from the dangers of living in the rarely "wild," often "war-torn," typically "horribly impoverished" areas of the world to which they are indigenous, animals in today's enlightened zoos, with their veterinarians and antibiotics, can now look forward to long lives and reproductive success.

Indeed, by a logic which has never really been challenged, propagation has become the final and apparently all-convincing register of both animal happiness and the importance of zoos. How often have we heard that a zoo has so effectively recreated an animal's natural home that, finally comfortable with its captivity, it has successfully bred? How often have we heard that the real reason zoos exist is to protect and conserve the world's animals? Despite the over-whelmingly obvious fact that zoos are created, maintained, and expanded for the pleasures of a human audience, ever since Hagenbeck's natural landscapes and immersion exhibits in which animals seemed free, defenders of zoos have been talking about animal happiness. Perhaps not surprisingly, though, if an animal does not conform to a now almost ubiquitous standard of propagational contentment, zoos have new technologies at their disposal to reassure a wary audience. The new Ark, we are told, is not simply filled with two of every kind being ushered safely into a better future world; in the new Ark animals can look forward to genetic immortality as cryogenically preserved gametes and tissue samples.

The enthusiasm with which zoo professionals have embraced such reproductive technologies as in vitro fertilizations, frozen-thawed embryo transfers, and nuclear transfers to "reproduce" particularly endangered or charismatic species such as elephants, pandas, great apes, and African wildcats suggests just how deeply the idea of the zoo as an Ark has resonated within the zoo world (see Loskutoff; and Goodrowe). At the beginning of the twentieth century, as concern grew over the disappearance of several notable species, no zoological garden sought to claim that it was, before anything else, a conservation organization. At the time, zoos were seen to fulfill such other more important or exigent goals as providing opportunities for scientific investigation, classrooms for children, and recreation for weary urban workers. It was fun to go to the zoo, or educational, or of scientific interest, and that was about it. A hundred years later, we are being told that zoos are less for people than for animals. Now when we hear people quietly protest (only loud enough to be heard by close confidants) that they can hardly see anything in the new exhibits, or when we hear the disappointment of people who have just discovered that their local zoo does not offer elephant rides anymore, we are all supposed to realize that these changes—the changes that make the Bronx Zoo into the Wildlife Conservation Society—have been made because the zoo should be a place for the care and protection of animals, not for the amusement of people. When we hear about the impressive Species Survival Programs (SSPs), in which accredited zoos work together to breed endangered animals, we are not supposed to trace their origin to the difficulties of obtaining new wild-caught specimens in a world of international laws and treaties designed to protect animals from commercial trade. Rather, we are expected to trace their origin to the genuine desire of zoos to scientifically assure the survival of a species in captivity with the hope that one day the animals may be returned to the wild from captive populations—something which has been successfully accomplished already with a handful of species. When we peer into multi-million-dollar immersion exhibits, we are supposed to understand that these exhibits exist primarily to make the animals happy. But could this really be true?

From the very beginning of experiments with immersion exhibits, it was clear that this type of exhibit was designed for the pleasure of the public. When people came to Hagenbeck's Park and saw the animals living in apparent freedom, they were ecstatic. But Hagenbeck's Park was not about restoring animals to their natural environments. It was better than that. As should be clear from a photograph of the Park's main panorama, the immersion exhibit was never really intended to trick people into believing they had stepped into a natural scene. Just as everyone visiting the "Congo Gorilla Forest" knows that they have not been miraculously transported to the west coast of Africa and that they are, in fact, visiting a zoo in the Bronx in New York City, Hagenbeck's goal was not accurate simulation. His goal—and that of all designers of immersion exhibits —was to convince people to suspend their disbelief long enough to accept what they saw before them as an alternative but believable scene. The goal of the immersion exhibit was and is to create a convincing verisimilitude. But to be convincing, it seems, the immersion exhibit must actually outdo nature. Compressed into small spaces, the better nature of the zoo makes real nature seem dull in comparison. The nature of the zoo suggests that there should be an animal—or better yet many animals—in every scene, and that one should only have to look hard enough to find them. But is all this good for the animals? Or, more broadly, does the attention lavished on a particular gorilla or pair of young pandas in Atlanta yield improved chances for their species?

The answer to this seemingly simple question is not easy. Consider, for example, the case of Keiko the Killer Whale, who starred in the movie *Free Willy* in 1993. Through the broad sentimentalization of Willy/Keiko, tens of millions of dollars continue to be raised and spent to return the whale to the wild. Keiko is a first-rate animal star and images from the Keiko-cam (one of the first of the increasingly popular zoo-cams continually posting pictures of zoo celebrities to the Internet) were downloaded by the hundreds of thousands while the whale was living at the Oregon Coast Aquarium. It seems reasonable to suggest that the story of Keiko has, indeed, made some people more aware of the difficult lives of large marine mammals in zoological gardens and aquariums. What is also completely clear, however, is that the major commercial aquariums have not only endured criticism surrounding the life of Keiko, they have actually become more eager to have whales in their collections, because people want to "see Willy."

The point is that elaborate new high-tech immersion habitats/enclosures/cages for primates and pandas and other animals—exhibits that make celebrities out of the animals and out of the zoo directors—seem only to generate a need for more spectacular exhibits and more spectacular animals. This is why "panda-mania" has been so frequently criticized by people who are interested in the preservation of pandas in China. There is a catch-22 in panda-mania. On one hand, by leasing pandas to American and European zoos for exorbitant fees, China gains needed cash. The money is at least partly used for panda research and protection; and, of course, for breeding more pandas for more zoos, and so on. In short, it is clear that the public interest in pandas does in the end contribute in some way to their protection and propagation in China. But is it accurate then to claim that those zoos which are paying millions of dollars to lease pandas are doing so for the sake of the pandas? The reason any zoo wants pandas—

11.12. "Bill's" first experience with grass; spider monkey. Milwaukee
County Zoo.
Photo courtesy of Mark Scheuber.

or gorillas, or koalas for that matter—is that the public wants to see those animals and will pay for the opportunity to do so. We have returned to the desires of the public having primacy. We have returned to Hagenbeck. We have returned to the invention of immersion exhibits.

Zoos exist because people find them interesting or relaxing or fun or educational places to go. Ever-more-realistic exhibits at zoos exist not because they tend to lessen the amount of stereotyped behavior seen in the animals (which they often do), nor even because animals often find contact with real soil and plants interesting and enjoyable (fig. 11.12). They exist because people have come to dislike looking at animals behind bars and in small glassed-in rooms and prefer exhibits in which animals appear to be living in nature. But the strange nature of zoos is such that it is clear to at least one primate keeper I know that her charges are more relaxed and content in their off-exhibit areas than in their high-tech immersion exhibit; so clear, in fact, that she will refer to the large habitat constructed for them as the place where the animals "work," and say that the animals enjoy returning to and relaxing in their smaller quarters for the night. Is it necessarily bad that animals "work" at the zoo? Work is a diversion, it breaks up the day, it can be mentally and physically stimulating, it can be like living in the wild. But this isn't "the wild," and it isn't even a replica of "the wild." It's a fantasy of "the wild" reinforced by nature television where at every turn the camera seems unbelievably ready—and the light implausibly perfect—to catch the most unimaginable shot. It is "the wild" that more and more people seem to think is some kind of "real thing." Unlike the actual Congo, where one can walk for days without seeing anything larger than an insect, in the new, better world of the immersion "Congo" life teems—at least the big life forms that attract so much human interest and attention. For life to teem in this way, for all the particularly fascinating species to be always present, happy, and breeding, it is essential that zoos take as complete control as possible of the future of rare animals. For the logic of the zoo to hold together, zoos must become the last, best hope of the world's threatened fauna.

No longer simply places of human amusement where strange animals are on display, the new(est) zoos, we are to conclude, provide ideal sanctuaries for animals. Just how appealing to the public this sort of argument has become is clear when one tries to comprehend the almost science-fictional world that zoos are beginning to imagine for themselves when they talk about using the most advanced reproductive technologies to bring animals back from the edge (and, with cloning, even from beyond the edge) of extinction. As Vicki Croke, a recent chronicler of zoos, has put it,

> Herds of elephants, countless rhinos, cheetahs, tigers and gorillas exist today in a state of suspended animation, thousands of the most precious species riding on a timeless plane into a mysterious future. They do not eat or drink. They are not aging. They are safe from disease. And each requires no more space than a plastic drinking straw. (165)

At several of our most important zoos today, tissue samples, eggs, sperm, and frozen embryos—the genetic diversity of at least those few large animals we love to see in zoos (although not the thousands of other threatened but insufficiently glamorous species)—are now safely protected for the future in carefully designed vessels bathed in frigid liquid nitrogen. As in most stories that have a happy ending, these "animals" in the zoo, heroic protagonists who are literally displayed in the most immersing of exhibits, truly will, we are assured, live happily ever after.

Notes

Portions of this essay are taken from my book *Savages and Beasts*.

1.   I use "better" here as the American zoo industry does; it separates out the vast majority of exotic animal collections in the U.S. from those "better" collections which are "accredited" by the professional American Association of Zoo-

logical Parks and Aquariums (AAZPA), also referred to as the American Zoo Association (AZA).

2. My thanks to Rory Browne for identifying the artist of this print.

3. "Don't Feed The Animals!" signs began to make their appearance at zoological gardens in the 1930s.

4. This guide was one of the many unofficial guides to the Gardens. The Society's "official guides" were published at irregular intervals from 1829 to 1857, when a standard format was adopted under the secretaryship of D. W. Mitchell.

5. John P. Haines to William Hornaday, 4 Sep. 1903. Incoming Correspondence, Director's Office, Archives of the New York Zoological Park, The Wildlife Conservation Society.

6. On the history of zoological gardens see especially Fisher; Hancocks; Hediger; Hoage and Deiss; Kisling; Knauer; Loisel; Mullan and Marvin; Peel; Ritvo; and Robbins. For the history of the firm of Carl Hagenbeck, see especially Dittrich and Rieke-Müller; Leutemann; Niemeyer; Pelc and Gretzschel; Reichenbach; Rothfels; Sokolowsky, *Carl Hagenbeck*; and Zukowsky.

7. In his recent essay on the history of "natural" exhibits in zoos, Jeffrey Hyson does an admirable job of denouncing the rhetoric of progress suggested by members of the zoo profession when they claim that "immersion exhibits" began in 1978 with the Woodland Park Zoo in Seattle. In response to designers' and directors' patting themselves on the back for their remarkable new creations, Hyson writes that the "triumphalist narrative is deeply flawed, in both its history and its conclusion. . . . [T]his same tale has been told again and again over the past century or more, as each new generation's directors and designers have proclaimed themselves more enlightened than their noble but misguided predecessors" (25). In the end, though, Hyson remains more sanguine than I can be about contemporary zoo designs. He concludes, "While the best work of today's zoo designers is impressive, exciting, and invaluable to our appreciation of wildlife, their confident environmentalism is challenged when viewed in the historical context of the planning and perception of zoos' 'natural' landscapes" (25).

8. All translations are my own, unless otherwise noted.

9. That Hagenbeck's moated enclosures seem relatively predictable at this point obscures how controversial they were when they were first created. Peter Chalmers Mitchell, the secretary of the London Zoological Society, dismissed Hagenbeck's exhibits from the start as "frankly theatrical scenery"; it was, of course, this scenery which would serve as the principal inspiration for Mitchell and J. P. Joass's 1914 designs for the Mappin Terraces at the London Zoo (see Mitchell).

10. Alexander Sokolowsky, one of the Park's scientific assistants in the 1920s, seems to fall naturally into the dual paradigms of Ark and Eden in speaking about the Park's main panorama: "There the viewer sees, in trustful unity, zebras, eland, gnus, and many other creatures, peacefully moving about each other unconcerned about each others' activities. A practiced animal observer, however, will quickly notice that the different species keep to their own, just like in Noah's Ark, where the pairs were brought together by Father Noah, or in the pictures of the animal paradise for which we can thank the imagination of medieval artists" (*Carl Hagenbeck* 48).

11. The quotation here from the English edition, *Beasts and Men*, is a straightforward translation from the unabridged 1908 original German *Von Tieren und Menschen*, 436–37.

12. Gorilla death is strongly connected to depression due to captivity in, for example, Sokolowsky's slim and somewhat enigmatic volume titled *Beobachtungen über die Psyche der Menschenaffen*.

13. The caption in the early English editions read simply "The Three Friends." The picture is discussed in the important work of Mullan and Marvin (85–87).

14. My thanks to Marcus Bullock for his help in clarifying the sense of the perplexing caption.

15. Through an examination of the plans of the New York Zoological Society at the Bronx Zoo and at the Jackson Hole Wildlife Park, Gregg Mitman has demonstrated just how profoundly ideas of the management of "nature" and "wildlife" could enter into conceptions of both zoos and "wild" populations of animals. My thanks to Chris Young for sharing this article with me.

Bibliography

Croke, Vicki. *The Modern Ark: The Story of Zoos: Past, Present and Future.* New York: Avon, 1997.

Dittrich, Lothar, and Annelore Rieke-Müller. *Carl Hagenbeck (1844–1913): Tierhandel und Schaustellungen im deutschen Kaiserreich.* Frankfurt am Main: Lang, 1998.

"Fast Facts." *Wildlife Conservation Society.* Accessed 30 Jan. 2002 <http://www.congogorillaforest.com/i-fastfacts.html>.

Fisher, James. *Zoos of the World.* London: Aldus, 1966.

Goodrowe, Karen L. "The Role of Genome Resource Banking in Wildlife Conservation Programs." *Communiqué* Feb. 2001: 13–14.

Guillery, Peter. *The Buildings of London Zoo.* London: Royal Commission on the Historical Monuments of England, 1993.

Hagenbeck, Carl. *Beasts and Men: Being Carl Hagenbeck's Experiences for Half a Century among Wild Animals.* Ed. and trans. Hugh S. R. Eliot and A. G. Thacker. London: Longmans, 1912.

———. *Von Tieren und Menschen: Erlebnisse und Erfahrungen.* Leipzig, 1908.

Hancocks, David. *Animals and Architecture.* New York: Praeger, 1971.

Hediger, Heini. *Man and Animal in the Zoo: Zoo Biology.* Trans. Gwynne Vevers and Winwood Reade. New York: Delacorte, 1969.

Hoage, R. J., and William A. Deiss, eds. *New Worlds, New Animals: From Menagerie to Zoological Park in the Nineteenth Century.* Baltimore: Johns Hopkins University Press, 1996.

Hyson, Jeffrey. "Jungles of Eden: The Design of American Zoos." *Environmentalism in Landscape Architecture.* Ed. Michel Conan. Washington, D.C.: Dumbarton Oaks Research Library and Collection, 2000. 23–44.

Katt, Friedrich. "Hagenbecks Tierparadies." *Zoologische Beobachter* 50 (1909): 370–72.

Kisling, Vernon N., Jr., ed. *Zoo and Aquarium History: Ancient Animal Collections to Zoological Gardens.* Boca Raton, Fla.: CRC, 2001.

Knauer, Friedrich. *Der Zoologische Garten: Entwicklungsgang, Anlage, und Betrieb unserer Tiergärten.* Leipzig: Theodore Thomas, [1907?].

Koebner, Linda. *ZooBook: The Evolution of Wildlife Conservation Centers.* New York: Forge, 1994.

Leutemann, Heinrich. *Lebensbeschreibung des Thierhändlers Carl Hagenbeck.* Hamburg, 1887.

Loisel, Gustav. *Histoire des menageries.* Paris: O. Doin, 1912.

Loskutoff, Naida. "Giving Nature a Helping Hand." *Communiqué* Feb. 2001: 4–6, 43.

Mitchell, Sir Peter Chalmers. *Centenary History of the Zoological Society of London.* London: Zoological Society of London, 1929.

Mitman, Gregg. "When Nature *Is* the Zoo: Vision and Power in the Art and Science of Natural History." *Osiris* 2nd ser. 11 (1996): 117–43.

Mullan, Bob, and Garry Marvin. *Zoo Culture.* London: Weidenfeld and Nicholson, 1987.

Niemeyer, Günter H. W. *Hagenbeck: Geschichte und Geschichten.* Hamburg: Hans Christians, 1972.

"Okapi Jungle." *Wildlife Conservation Society.* Accessed 30 Jan. 2002 <http://www.congogorillaforest.com/vts-okapijungle.html>.

Peel, C. V. A. *The Zoological Gardens of Europe: Their History and Chief Features.* London: Robinson, 1903.

Pelc, Ortwin, and Matthias Gretzschel. *Hagenbeck: Tiere, Menschen, Illusionen.* Hamburg: Hamburger Abendblatt, 1998.

Priemel, Kurt. "Handelstierpark und zoologische Gärten." *Frankfurter Zeitung und Handelsblatt* 24 Apr. 1909: n.p.

Reichenbach, Herman. "Carl Hagenbeck's Tierpark and Modern Zoological Gardens." *Journal of the Society for the Bibliography of Natural History* 9.4 (1980): 573–85.

———. "A Tale of Two Zoos: The Hamburg Zoological Garden and Carl Hagenbeck's Tierpark." Hoage and Deiss, *New Worlds,* 51–62.

Rilke, Rainer Maria. "The Panther, Jardin des Plantes, Paris." *Neue Gedichte.* 1907. Trans. Stephen Cohn. Manchester: Carcanet, 1992. 61.

Ritvo, Harriet. *The Animal Estate: The English and Other Creatures in the Victorian Age.* Cambridge, Mass.: Harvard University Press, 1987.

Robbins, Louise E. *Elephant Slaves and Pampered Parrots: Exotic Animals in Eighteenth-Century Paris.* Baltimore: Johns Hopkins University Press, 2002.

Rothfels, Nigel. *Savages and Beasts: The Birth of the Modern Zoo.* Baltimore: Johns Hopkins University Press, 2002.

Scherren, Henry. *The Zoological Society of London: A Sketch of Its Foundation and Development.* London: Cassell, 1905.

Sokolowsky, Alexander. *Beobachtungen über die Psyche der Menschenaffen.* Frankfurt am Main: Neuer Frankfurter Verlag, 1908.

———. *Carl Hagenbeck und sein Werk.* Leipzig: Haberland, 1928.

Tudge, Colin. *Last Animals at the Zoo: How Mass Extinction Can Be Stopped.* Washington, D.C.: Island, 1992.

"Welcome to WCS's Congo Gorilla Forest Online!" *Wildlife Conservation Society.* Accessed 4 Feb. 2002 <http://www.congogorillaforest.com/>.

The Zoological Gardens: A Description of the Gardens and Menageries of the Zoological Society. A Handbook for Visitors. London: Clarke, [ca. 1861–62].

Zukowsky, Ludwig. *Carl Hagenbecks Reich: Ein deutsches Tierparadies.* Berlin: Volksband der Bücherfreunde, 1929.

Contributors

Steve Baker, Reader in Contemporary Visual Culture at the University of Central Lancashire, U.K., is the author of *The Postmodern Animal* (2000) and *Picturing the Beast: Animals, Identity, and Representation* (1993, 2001). He is also guest editor of *Society and Animals* 9.3 (2001), a special issue on "The Representation of Animals."

Marcus Bullock is Professor of English at the University of Wisconsin–Milwaukee. He is the author of *The Violent Eye: Ernst Jünger's Visions and Revisions on the European Right* (1992) and the editor, with Michael Jennings, of *Walter Benjamin: Selected Writings,* vol. 1 (1996).

Jane Desmond is Associate Professor of American Studies and Co-director of the International Forum for U.S. Studies at the University of Iowa. She is the author of *Staging Tourism: Bodies on Display from Waikiki to Sea World* (1999) and the editor of *Dancing Desires: Choreographing Sexuality* (2001) and *Meaning in Motion: New Cultural Studies of Dance* (1997).

Erica Fudge is Senior Lecturer in the School of Humanities and Cultural Studies at Middlesex University. She is the author of *Perceiving Animals: Humans and Beasts in Early Modern English Culture* (1999) and co-editor, with Ruth Gilbert and S. J. Wiseman, of *At the Borders of the Human: Beasts, Bodies, and Natural Philosophy in the Early Modern Period* (1999). She is currently editing a collection of essays, *Renaissance Beasts,* that will be published by Illinois University Press in 2002.

Andrew C. Isenberg is Assistant Professor of History at Princeton University. He is the author of *The Destruction of the Bison: An Environmental History, 1750–1920* (2000), in addition to articles in *Environmental History* and *Journal of the Early Republic.*

Kathleen Kete, Associate Professor of History at Trinity College, is the author of *The Beast in the Boudoir: Petkeeping in Nineteenth-Century Paris* (1994). Her essays and reviews have appeared in *Representations, Signs,* and the *Journal of Modern History.* She is currently writing a book on ambition in postrevolutionary France.

Akira Mizuta Lippit is Associate Professor of Film Studies and Critical Theory in the Program in Film and Visual Studies at the University of California, Irvine. He is the author of *Electric Animal: Toward a Rhetoric of Wildlife* (2000). His work has appeared in *Afterimage, Assemblage, MLN, Qui Parle,* and *Women and Performance.*

Teresa Mangum is Associate Professor of English at the University of Iowa and the author of *Married, Middlebrow, and Militant: Sarah Grand and the New Woman Novel* (1998). Her recent publications include essays in *Figuring Age: Women, Bodies, Generations* (ed. Kathleen Woodward, 1999) and *A Companion to Victorian Literature and Culture* (ed. Herbert F. Tucker, 1999). She is currently at work on a book to be titled *The Victorian Invention of Old Age.*

Garry Marvin is Senior Lecturer in Social Anthropology at the University of Surrey Roehampton. He is the author of *Bullfight* (1988, 1994) and, with Bob Mullan, *Zoo Culture* (1987, 1999), and is at present writing a book on English foxhunting. The film *The Hunt,* based on his research, received the Prix d'Italia for the best cultural documentary on European television in 1998.

Susan McHugh is Marion L. Brittain Fellow of Writing in the School of Literature, Communication, and Culture at the Georgia Institute of Technology. She is the author of essays in *Critical Inquiry* and *South Atlantic Review,* as well as reviews in *Modern Fiction Studies.* She is currently working on a book to be titled *Animal Cultures: Domesticated Animals and Visual Narrative.*

Nigel Rothfels is an independent scholar and Director of the Edison Initiative at the University of Wisconsin–Milwaukee. He is the author of *Savages and Beasts: The Birth of the Modern Zoo* (2002), and is currently writing a cultural history of the elephant.

Index

Italicized page numbers indicate illustrations.

68, 74, 95, 186–187; in foxhunting, 147–148; of Ernest Hemingway, 108, 109–110, 116, 120; of Franz Kafka, 110, 111–112; of D. H. Lawrence, 108, 109–110, 116, 120; and modernism, 124; of Ernest Thompson Seton, 50–51, 59–60; Victorians' use of, 44

Antisemitism, 27, 30. *See also* Nazism

Art, postmodern, 67, 68–70, 74, 81

Artaud, Antonin, 128

Ashton, Edwina, 68–69, 79, 82, 84; *Bear-Faced Monologue,* 68, *69,* 92; *Frog,* 68, 90; *Sheep,* 68, 90; *Slug Circus,* 68

ASPCA. *See* American Society for the Prevention of Cruelty to Animals

Atkins, Charles Henry (Speedy), 166, 167

Authenticity, 49, 60n3, 69, 161. *See also* Realism

Babe (film), 159, 170. *See also* Animation

Bacon, Francis, 119–123

Baker, Steve, 168

Bambi, 48, 49, 53, 55. *See also* Animation; Disney, Walt; Nature; Salten, Felix

Bands of Mercy. *See* Animal Defender

Barraud, Francis: *His Master's Voice,* 38

Barthes, Roland: and animation, 128; *Camera Lucida,* 119, 120, 122

Baseman, Jordan, 168. *See also* Art, postmodern; Taxidermy

Bataille, George, 125

Baudrillard, Jean, 86

Baudry, Jean-Louis, 126, 133n13

Bazin, André, 123, 131n4

Beauty. *See* Aesthetics

Beer, Gillian, 43

Bell, Alexander Graham, 125

Benchley, Peter, vii–xi, xii

Benjamin, Walter, 3, 101

Berger, John, 121, 131n3

Bernard, Claude: *Introduction to the Study of Experimental Medicine,* 27, 29

Beuys, Joseph: *Coyote,* 69, 85

Blackbourn, David, 24

Blackwood's Edinburgh Magazine: "My Old Dog and I," 41–42

Blanca (wolf), 51, 59

Blood sports, 21, 22

Body, the: of the animal, 99, 159, 160; in animation, 124; and becoming-animal, 75, 77–79, 82; and the carcass, 119–120; and cinema, 129; and identification, 102, 118; without organs, 84, 85–86, 88 (*see also* Deleuze, Gilles, and Felix Guattari); in taxidermy, 164–170; and transference, 127–128

Breath, metaphor of, 74, 75, 78, 91. *See also* Cixous, Hélène; Orozco, Gabriel

Breeding: and aesthetics, 187–188; and class, 105, 188–189, 190; of dogs, 36, 181–189; of foxhounds, 146–150

British Kennel Club, 36

Brontë, Emily, 40

Bruford, W. H., 24

Budiansky, Stephen, 36

Bullbaiting, 21

Bull-running, 21, 22

Burroughs, John, 50

Burton, Decimus, 206

Camus, Albert: *The Plague,* 100; *The Stranger,* 100

Capitalism, 56; and hunting, 24, 26

Captivity, 211–217, *215. See also* Display, animal; Zoos

Carlo (dog), 36

Cartmill, Matt: *A View to a Death in the Morning,* 150–151

Cinema, and animal death, 123–124, 125, 128–131

Endangered Species Act, 49, 56, 57
Environmentalism, 53–54, 56; and
  environmental history, 48
Erlich, Terry, 162
The Eternal Jew (film), 29
Ethics, 49, 56, 162; and morality, 54,
  55, 144, 154, 168
Euthanasia, veterinary, 37, 41, 44, 45
Evans, Nicholas: *The Loop*, 59, 60

Fabre, Jan: *A Consilience*, 90, *91*
Feminism, 27, 28, 29, 31, 82. *See also*
  Gender
Fischinger, Oskar, 130
Fitzgerald, William G.: "Dandy
  Dogs," 36–37
Flores, Nona C.: *Animals in the
  Middle Ages*, 8
Form, in art, 68, 84–89, 92, 101, 103,
  107, 112
Fox, Rodney, viii
Foxes: and aesthetics in foxhunting,
  147, 152–153; and anthropomor-
  phism in foxhunting, 147–148;
  and foxhounds, 146–150; and fox-
  hunting, 139–155; and the role
  of death during the Hunt, 140,
  149–152, 154; in representation,
  143–146
Free Willy (film), 218
Freedom: 110–111, 114, 210–212;
  and liberation, 20, 21, 30, 32, 33
Freud, Sigmund, 125, 126, 127, 129–
  130. *See also* Abraham, Nicolas,
  and Maria Torok; Transference

Game, 20–21, 23, 53, 57, 145, 151;
  and the Game Law of 1671, 24, 25;
  and the Game Reform Law of
  1831, 25; and protectionism, 24.
  See also Foxes; Hay, Douglas; Hunting; Munsche, P. B.; Predation;
  Wolves
Garber, Marjorie: *Dog Love*, 182,
  185, 186, 189, 194, 196

Gender, 43; and hunting, 24. *See also*
  Feminism
Goodall, Jane, xi
Great White, Deep Trouble (film), xii
*Great White Shark: Truth behind the
  Legend* (film), viii, ix–x
Greenaway, Peter: *The Falls* (film), 90
Grégoire, Henri, 31
Guattari, Felix. *See* Deleuze, Gilles,
  and Felix Guattari

Haecceity, 78, 85–86. *See also*
  Deleuze, Gilles, and Felix Guattari
Hagenbeck, Carl, 208–217
Haines, John P., 207–208
Harrison, Brian: *Peaceable Kingdom*,
  19, 37
Hay, Douglas: "Poaching and the
  Game Laws on Cannock Chase,"
  19, 24–25
Hayden, Jacqueline: *The Lion in Win-
  ter*, 203
Hearne, Vicki: *Adam's Task: Calling
  Animals by Name*, 19, 185
Heidegger, Martin, 92
Hemingway, Ernest: and anthropo-
  morphism, 108, 109–110; *The Old
  Man and the Sea*, 102–105, 108–
  110, 113, 114, 115
Hirst, Damien, 81–82, 92, 168; *Some
  Comfort Gained*, 81, *81*
Holism, 8
Horkheimer, Max: *Dialectic of Enlightenment*, 113, 114, 117
Human, the: and becoming-animal,
  67, 68, 74, 77–82, 90, 93–96; and
  the history of animals, 5–10, 20,
  21, 60; and humanism, 8–9, 26, 30,
  38–42, 50, 51, 67, 88, 93, 99, 117,
  121–124; vs. the animal, 49, 102–
  111, 129, 155, 166, 175, 187
Hunting: and capitalism, 24, 26; and
  class, 20–25, 32, 54–55, 140–155;
  formal, 32, 140–155; of foxes, 39–
  55; and gender, 24; as sport, 150–

Phineas, Charles: "Household Pets and Urban Alienation," 4–5, 6, 19

Photography, 119–123, 128. *See also* Barthes, Roland; Cinema, and animal death; Noelker, Frank

Poaching, 23, 144–145; legislation against, 48, 54. *See also* Game

Porter, Roy, 6

Predation, 49, 54, 55, 140, 144, 149, 150, 151; and predator eradication, 56, 57–58

Preservationism, 48, 113

Priemel, Kurt: on Carl Hagenbeck's Animal Park, 210

Prins, Gwyn, 5

Protectionism, 49, 56. *See also* Animal protection

Protectorate's ordinance of 1654, 21

Punch (magazine), 40

Puritans, and animal protection, 20, 21

Raffles, Thomas Stamford, and the Zoological Gardens of London, 205

Ramee, Marie de la. *See* Ouida

Rank, Otto, 127

Realism, 43, 74, 219; in taxidermy 159, 161–162, 163–164, 170, 173, 174–175

Regan, Tom: *The Case for Animal Rights,* 19, 32

Reification, 189–190

Reliquary, 165

Rights: animal, 4, 7, 19, 30–31, 32; customary, 55; human, 31–32; hunting, 24. *See also* Regan, Tom; Singer, Peter

Rilke, Rainer Maria, 211; "Archaic Torso of Apollo," 101, 102, 115

Ritual, hunting as, 140. *See also* Hunting

Ritvo, Harriet: *The Animal Estate: The English and Other Creatures in the Victorian Age,* 9, 10, 19, 25, 44, 188

Riviere, Briton: *Requiescat,* 38

Romanes, George, 43

Romanticism, 58, 59, 60; and hunting, 23; and nature, 26

Roof, Judith, 189, 194

Roosevelt, Theodore, 52

Royal Society for the Prevention of Cruelty to Animals (RSPCA), 22, 25, 26, 28, 29, 37

RSPCA. *See* Royal Society for the Prevention of Cruelty to Animals

Ryder, Richard, 4

Salisbury, Joyce E.: *The Medieval World of Nature,* 8

Salten, Felix: *Bambi: A Forest Life,* 55

Sanders, Clinton R., 44–45

Saunders, Marshall: *Beautiful Joe: An Autobiography,* 37, 40–41

Sax, Boria: "Understanding Nazi Animal Protection and the Holocaust," 30

Schama, Simon: *Citizens: A Chronicle of the French Revolution,* 23

Schiff, Moritz, 28

Schneemann, Carolee, 75, 78, 79, 82, 84; "Animal," 73; *Infinity Kisses,* 71, *72,* 73, *73; Vespers Pool,* 73–74

Schrodinger, Erwin, 125, 132–133n10

Schwartz, Marie-Espérance von: *Gemma, or Virtue and Vice,* 28, 29

Schwartz, Vanessa, 164

Scott, Sir Walter: *Old Mortality,* 39–40

Sebeok, Thomas, 125

Secord, William: *Dog Painting, 1840–1940: A Social History of the Dog in Art,* 37–38

Seeing, 115, 118; and eating, 122; and observing, 102–109

Self, Will: *Great Apes,* 90

Seton, Ernest Thompson, 49, 52, 55, 58, 59, 60; *Wild Animals I Have Known,* 50–51

Sewell, Anna: *Black Beauty,* 27, 37

Shapcott, Jo: "The Mad Cow Talks

Williams, David, 90
Williams, Greg, 71
Wise, John Sargeant, 40
Wodehouse, P. G., 25
Wollstonecraft, Mary, 30, 31
Woloch, Isser, 23
Wolves, 49, 51–52, 55, 58–60, 79, *80,* 91; hunting of, 23–24, 51–55, 59. *See also* Blanca; Lobo; London, Jack; McIntyre, Rick; Predation
Wright, Susan, 181
Writing, 74–75, 78, 95

Xenophon: *On Hunting,* 146

Yellowstone National Forest, 48, 53, 57, 58, 59. *See also* National Forest Service
Yosemite National Forest, 48. *See also* National Forest Service

Zooerasty, 121
Zoopraxiscope, 123
Zoos, 93, 199, 201, 202, *203;* and the bourgeoisie, 207, 208; history of, 205–217, *206, 210, 219*
Zukowsky, Ludwig, on Carl Hagenbeck's Animal Park, 212